Recording Music on Location

Recording Music on Location

Capturing the Live Performance

Bruce Bartlett
Jenny Bartlett

Focal Press
Taylor & Francis Group

NEW YORK AND LONDON

First published 2007 by Focal Press
70 Blanchard Road, Suite 402, Burlington, MA 01803

Simultaneously published in the UK by Focal Press
2 Park Square, Milton Park, Abingdon, Oxon OX14 4RN

Focal Press is an imprint of the Taylor & Francis Group, an informa business

Notices
Practitioners and researchers must always rely on their own experience and knowledge in evaluating and using any information, methods, compounds, or experiments described herein.

Product or corporate names may be trademarks or registered trademarks, and are used only for identification and explanation without intent to infringe.

Library of Congress Cataloging-in-Publication Data
Application submitted

ISBN 13: 978-0-240-80891-8 (pbk)
ISBN 13: 978-0-240-80943-4 (CD-ROM)

This book is fondly dedicated to the memory of Mom, Dad, and Tom Lininger.

CONTENTS

2 Recording Techniques from Simple to Complex   29

3 Before the Session: Planning   45

9 Stereo Recording Procedures   121

10 Surround-Sound Miking Techniques   141

11 Troubleshooting Stereo Sound   157

12 Stereo, Surround, and Binaural Microphones and Accessories   165

A Stereo Imaging Theory   181

B Specific Free-Field Stereo Microphone Techniques   209

Contents

PREFACE

One listen to "Do You Feel Like We Do" by Peter Frampton, and you'll know why live recordings can be so thrilling.

Perhaps the most exciting type of recording is done with the musicians playing "live" in a club or concert hall. Many bands want to be recorded in concert because they feel that's when they play best. They take chances and surprise the audience. Your job is to capture that performance and bring it back alive.

Without a doubt, remote recording is exhilarating. The musicians, responding to the audience's energy, often put on a stellar performance. You have only one chance to get it recorded and it must be done right. You're working on the edge. But by the end of the night, when everything has gone as planned, it's a great feeling.

This book, *Recording Music On Location*, will help you do it right. It is the first book to focus exclusively on the special techniques used for recording outside the studio. It covers the unique requirements for capturing sound in a room or hall where the music is performed.

Whether you want to record an orchestra in a concert hall, a jazz combo in an auditorium, a rock band in a club, or a touring band on the road, this book will offer the practical advice to help you do it. The new breed of compact mixers, flash-memory recorders, digital audio workstations, and multitrack recorders has made going on location easier than ever. This book was written to help you take advantage of these new tools.

Recording Music On Location is intended for recording engineers, live sound engineers, record producers, musicians, hobby recordists, concert tapers, and podcasters—anyone who wants to know more about remote recording.

Maybe you're a musician who wants to record your band. If the band is too big for your home studio, or if noise is a problem there, you can go out to a venue and record the band in a live performance. With less cost than it takes to record in a professional studio, you can record a show and produce a live CD. This demo recording can be used to get gigs. Some bands start with live-recorded tracks, and then use them in the studio as a basis for developing complete productions.

Recording Music On Location is divided into two main parts: (1) popular music recording and (2) classical music recording. The recording styles for these types of music are quite different. Let's look at Part 1 first.

Part 1: Popular Music Recording (Rock, country, jazz, folk, R&B, gospel, Christian, and so on)

Starting off Part 1, Chapter 1 offers an overview of audio gear for recording pop music on location, both for two-track (stereo) and multitrack recording.

There are many ways to record live pop music, from simple to complex. Chapter 2 walks you through each method. You'll also learn how to interface with the sound-reinforcement (PA) system while making a multitrack recording.

Chapter 3 helps you plan a live multitrack recording session. Listing the equipment you need, and how you will record with it, will make the actual recording a lot easier and give you a better result. Based on my experience as an on-location recording engineer, this chapter also offers tips for easier setup. Here you'll find shortcuts to make your job go smoother.

In Chapter 4 we go over the procedures at the actual multitrack recording session: connecting to power, running cables, miking, console setup, and so on. Chapter 5 suggests ways to mix and edit a multitrack recording of a gig or concert.

Finally, Chapter 6 describes a real-world recording project: recording a blues band in a club.

The *Frampton Comes Alive!* audio CD by Peter Frampton was originally released in 1976, produced by Peter Frampton, engineered by Eddie Kramer and Chris Kimsey. For other great live recordings, check out www.allaboutjazz.com/php/article.php?id=14757.

Part 2: Classical Music Recording (Orchestra, string quartet, pipe organ, choir, soloist)

With popular music, it's common to use multiple close mics and multitrack recorders. But with classical music, stereo mic techniques are the norm. There are many ways to make true-stereo recordings, and Part 2 covers them all. It offers a clear, practical explanation of stereo miking theory, along with specific techniques, procedures, and hardware.

True-stereo microphone techniques use two or three microphones to capture the overall sound of the music and the concert hall. The stereo recording made from these microphones is usually reproduced over two speakers. Ideally, the goal is to produce a believable illusion of the musical ensemble and the concert hall in a solid, or three-dimensional, way.

For example, an orchestra might be recorded with two microphones and played back over two speakers. You would hear sonic images of the instruments in various locations between the stereo pair of speakers.

These image locations—left to right, front to back—correspond to the instrument locations during the recording session. In addition, the concert hall acoustics are reproduced with a pleasing spaciousness. The result can be a beautiful, realistic re-creation of the original event—or even an improvement on it.

Part 2, Chapter 7, starts by demystifying microphone polar patterns (directional pickup patterns), which are key to knowing which mics to use to create the effect you want. This is followed in Chapter 8 by an overview of the most common stereo microphone techniques.

Next, Chapter 9 leads you through the procedures in a classical music recording session: where to record, where to place the mics, recording tips, and so on. Chapter 10 covers several techniques for surround-sound miking. In a surround recording of classical music, we usually hear the orchestra up front, and we hear the concert hall ambience from all around. Special mic techniques have been developed for capturing this surround effect.

Also included are a troubleshooting guide for stereo sound and a listing of stereo, surround, and binaural mics and accessories. A glossary explains the technical terms in the book.

The appendices are the most academic sections. They are intended for audio engineers who want a deeper understanding of stereo and stereo mic techniques, or who want to create their own techniques.

Appendix A covers stereo imaging theory in detail: how we hear where sounds are coming from, how we localize "images" of musical instruments between loudspeakers, and how mic techniques create images in various locations. You'll learn how to configure stereo arrays to achieve various stereo effects.

Specific microphone techniques (such as XY, MS (mid–side), Blumlein, ORTF, OSS, SASS) are explained next: their characteristics, stereo effects, benefits, and drawbacks. Appendix B is devoted to free-field methods; Appendix C to boundary methods; and binaural techniques are covered in Appendix D.

I hope you enjoy the thrill of live recording as much as I do.

Part 1
Popular Music Recording
(Rock, country, jazz, folk, R&B, gospel, Christian, and so on)

1

GEAR FOR LIVE RECORDING

Whether you are a musician, concert taper, live sound engineer, or studio engineer, you'll find helpful information on live recording in this book. This chapter is an overview of the necessary equipment.

Your simplest option is to record live to a portable stereo recorder. The process is easy and the required gear costs under $600. A stereo recording may not offer the sound quality of a professional multitrack recording. But if you record how the band sounds from a seat in the audience, that may be good enough—especially if the recording is just for yourself or your friends.

Audio professionals can use a mobile recording rig with a multitrack hard-drive (HD) recorder or laptop computer. This setup is convenient and provides excellent sound for about $2000 and up. We'll look at the pros and cons of these options in a minute.

This chapter offers a basic survey of equipment for recording music on location. For a deeper understanding of recording technology, we suggest our book, *Practical Recording Techniques*, fourth edition, published by Focal Press.

Stereo Systems versus Multitrack Systems

You can make live recordings with a stereo recorder or a multitrack recorder. Basically, a stereo recording system uses two mics (or a stereo mic) plugged into a portable stereo recorder. The mics pick up the group as a whole from several feet away, and the mic signals are recorded. A multitrack system uses several mics, each close to an instrument or singer. The mic signals are amplified and sent to a multitrack recorder. One track might be a recording of the lead vocal, another track might be the sax, another the kick drum, and so on. You mix the tracks back in the studio.

Stereo recording is easy and cheap, and it captures the sound as heard in the audience (including the room reverberation and background noise). You could call it a "documentary" or "audio snapshot" recording. The multitrack approach is more challenging and expensive, but it offers a cleaner, more commercial sound, probably with a well-balanced mix. It's the most common method used by professional recording engineers to record live pop music.

A stereo recording can sound very good if no PA system is in use— but most bands use a PA. When you record the band you're also recording the sound of the PA speakers. Thus, the mix or balance you get depends on the PA engineer's skill.

The first half of this chapter focuses on stereo recording systems, while the second half covers multitrack systems.

Stereo Recording Systems

Figure 1-1 shows the parts of a typical stereo recording system using microphones. Placed several feet from the performers, the mics pick up the group, room sound reflections, and any background noise. The sound and signals move or flow from start to finish (left to right in Figure 1-1).

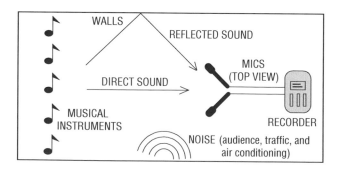

Figure 1-1 Signal flow in a typical stereo recording system.

This is the signal flow shown in Figure 1-1:

1. Musical instruments produce sound.
2. Background noise and room reverberation add to the musicians' sound.
3. Microphones pick up the total sound and change it into electrical signals.
4. Mic choice and placement affect the tone quality (bass and treble), the stereo effect, and the amount of background noise and room reverberation that are picked up.
5. Mic cables carry the mic signals to the recorder. Some mics plug directly into the recorder.
6. The recorder makes a stereo recording of the left- and right-mic signals.

If you can record off the PA mixing board, all you need is a portable stereo recorder and cables.

Equipment for Stereo Recording

Let's describe in detail the gear you need to make a simple stereo recording.

Microphones

A microphone changes sound into an electrical signal. Classified by how that is done, there are three types of mics for recording: condenser, dynamic, and ribbon.

Condenser, Dynamic, and Ribbon Types

Condenser mics typically give a clear, detailed, natural sound. They are the preferred choice for stereo recording. Condenser mics require a power supply to work, explained later under the heading "Mic Connectors, Powering, and Cables."

Dynamic (moving-coil) mics work without any power supply. They are rugged and reliable. Most dynamic mics do not sound as clear and natural as condensers and are less sensitive, so dynamics are seldom used for stereo recording.

A *ribbon* mic provides a smooth sound that many people prefer, and it works without power, but it's delicate and expensive.

Sound Pickup Patterns (Polar Patterns)

Microphones also differ in the way they respond to sounds coming from different directions:

- An *omnidirectional* (*omni*) mic picks up sound equally well in all directions.
- A *unidirectional* mic picks up sound best in front of the microphone. It partly rejects sounds to the sides and rear of the mic. Three types of unidirectional mic are *cardioid*, *supercardioid*, and *hypercardioid*. Each has a progressively narrower pickup pattern.
- A *bidirectional* (figure-eight) mic picks up best in two directions: in front of and behind the microphone. Most ribbon mics have a bidirectional polar pattern. Mics with this pattern are used in the Blumlein stereo technique, described in Chapter 8. Figure 7-1 in Chapter 7 shows various polar patterns.

Which mic pattern is right for your needs? Choose omni mics when you need all-around pickup, extra deep bass, less handling noise and wind noise, or binaural (headworn) miking for headphone playback. Choose cardioid mics when you need sharp stereo imaging, rejection of room reverberation, and rejection of background noise.

Mic Connectors, Powering, and Cables

As shown in Figure 1-2, mics come with either an XLR (3-pin) connector or a phone plug (called a "jack" plug outside the US). Most condenser mics with an XLR connector are powered by 12–48 V *phantom power*. This powering can be supplied by a phantom power supply, mic preamp, recorder, or mixer. Condenser mics with a phone plug (jack plug) either use an internal battery, or they receive *plug-in power* (3–10 V DC) from a mini mic preamp or recorder. Some mics can be powered by a separate *battery module*, which helps the mic pick up loud sound sources with less distortion (increased dynamic range).

A mic with an XLR connector has what's called a "low-impedance, balanced" output. Such a mic can be used with very long mic cables without picking up hum or losing treble. A mic with a phone plug (jack plug outside the US) comes with a short, permanently attached cable or no cable. This type of mic has an unbalanced output that is low-to-medium impedance.

What if your mics have XLR connectors, but your recorder or mic preamp has one or two phone jacks (sockets outside the United States)? You'll need an adapter cable, shown in Figure 1-3.

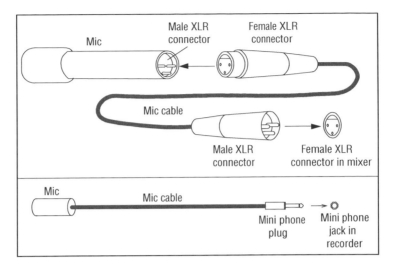

Figure 1-2 (Top): Male and female XLR connectors. (Bottom): Phone plug and phone jack connectors (jack plug and socket connectors outside the US).

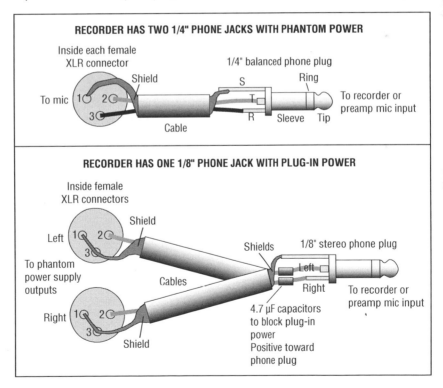

Figure 1-3 XLR-to-phone adapters (XLR-to-jack adapters outside the US).

Cheap 1/8-inch phone plugs (3.5 mm jack plugs) with thin gold plating are actually less reliable than plugs with nickel plating. Thin gold plating wears off quickly, exposing a brass surface that makes poor contact.

Special-Purpose Mics

A *stereo mic* has two mic capsules in the same housing for convenient stereo recording. A *mini stereo mic* plugs directly into some portable digital recorders. Mini stereo mics that use cardioid mic capsules tend to have less bass and more noise (hiss) than larger stereo mics. A headworn *binaural mic* has two miniature omni condenser mics that you wear in or near your ears; you play back the recording on headphones. Chapter 12 lists all these types of microphones.

You can make yourself a decent stereo or binaural mic for experimenting. Purchase some Panasonic WM-61B102B omni mic capsules from www.Digikey.com for $1.85 each. Get an adapter cable with a 1/8-inch stereo plug (3.5 mm stereo jack plug outside the US) and two RCA (phono) plugs. Cut off the two RCA (phono) plugs and solder the wires to the mics. If you want mics that are ruggeder, of higher quality, and better looking, check out the commercial mics listed in Chapter 12.

Microphone Mounting Styles

Microphones also can be classified by the way they mount onto objects:

- A stand-mounted stereo mic attaches to a mic stand, which provides the most secure and stable mounting. A stereo pair of mics can mount on a stereo bar (stereo mic adapter), which holds two mics on a single mic stand. However, mic stands might be too large to be acceptable in certain venues, and they are a hassle to carry.
- Plug-in mics plug into a mic preamp or portable recorder. No mic stand is needed.
- "Goosenoose" stereo mics are worn around the neck.
- Clip-on mics can be clipped to a shirt at the shoulders or to eyeglass earpieces.
- Headband-mounted mics are attached to a headband. Some headband products have "street" styling.
- Handheld mics have a handgrip. Watch out for rubbing noises if you use a handheld mic.
- Desktop mics sit a few inches above a desk or a table, so they might pick up an unnatural, filtered sound due to surface-sound reflections.

- Boundary mics eliminate this problem by mounting directly on surfaces.

Mic Specs

When you shop for a mic, consider these other specifications on the mic data sheet:

- *Signal-to-noise (S/N) ratio*: 67 dB at 1 Pa (Pascal) is fair; 74 dB is very good; 84 is excellent. But 67 dB is good enough if you are recording loud rock concerts.
- *Frequency response range*: 100 Hz to 15 kHz is fair; 50 Hz to 18 kHz is very good; 20 Hz to 20 kHz is excellent.
- *Frequency response tolerance*: ±6 dB is fair; ±3 dB is very good; ±1 dB is excellent.
- *Maximum sound pressure level*: 100 dB is fair; 110 dB is very good; 120 dB is excellent (high enough for rock concerts).
- *Size*: Small-diameter microphones (under 1/2 inch) tend to be relatively noisy, but this may not be a problem if you are recording loud music. Omni mics of any size can have excellent bass. They pick up deeper bass than small cardioid mics, which sound "thin" by comparison.
- *Accessories*: A foam windscreen for recording outdoors is a handy accessory. If your mic lacks a windscreen, you can purchase one from Radio Shack (or a music store) and cut it to fit. A stereo bar or stereo mic adapter mounts a pair of mics on a single mic stand for convenient stereo miking.

Portable Stereo Recorder

Having covered mics for stereo recording, let's move on to the recording device itself. We'll look at five different types of portable stereo recorder: a flash-memory recorder, flash-memory recording system, MiniDisc recorder, digital audiotape (DAT) recorder, and a laptop computer with recording software.

Flash-Memory Recorder

A flash-memory recorder (Figures 1-4 and 1-5) is a portable digital recorder with no moving parts. Also called a *solid-state recorder*, it records into a flash-memory card such as a Compact Flash or Secure Digital (SD) card. A 2 GB card, which records 2 hours of 24-bit/44.1 kHz wave audio

Figure 1-4 M-Audio Microtrack 2496, an example of a stereo flash-memory recorder.

Figure 1-5 Tascam HD-P2, an example of a stereo flash-memory recorder.

files, costs about $65. Flash-memory recorders can record MP3 or uncompressed PCM wave files (which are CD quality or better).

These recorders have a number of features to consider. Power comes from replaceable or rechargeable batteries. Available mic connectors are XLR, 1/4-inch phone (6.35 mm socket), or 1/8-inch phone (3.5 mm socket), with or without 48V phantom power or plug-in power. Some units come with built-in or plug-in stereo microphones. Prices range from $400 to $2000.

After making a recording, you connect the USB (Universal Serial Bus) port in the recorder to the USB port in a computer. The recorder shows up as a storage device on your computer screen. You drag-and-drop the recorded sound files to the computer's HD for editing and CD burning. The files transfer in a few minutes. Then the flash-memory card is empty, free to make more recordings.

Nearly all flash-memory recorders include a mic-gain switch to accommodate both quiet and loud sound sources. Low gain or low amplification (0–15 dB) is for recording loud sounds (rock concerts); medium gain (25 dB) is for recording medium sounds (acoustic music, lectures, or rehearsals); and high gain (50 dB) is for recording quiet sounds (nature, quiet talking). Most recorders have AGC (automatic-gain control), which sets the recording level automatically depending on how loud the sound is. Some units include a limiter to prevent recording above 0 dB level, which otherwise would cause distortion.

Some examples of flash-memory recorders are the Sony PCM-D1; Marantz PMD660, PMD670, and PMD671; Core Sound PDAudio® System, AEQ PAW-120, Edirol R-1 and R-09, M-Audio MicroTrack 2496, Nagra Ares-M, PocketRec software, Sound Devices 722/744T, Fostex FR-2, Roland CD2, Mayah Communications Flashman, and Tascam HD-P2.

Flash-Memory Recording System

A lower-cost alternative to a flash recorder is a flash-memory recording system made of several components (Figure 1-6(a)). This is the signal flow from start to finish (left to right):

1. A **stereo mic** (or mic pair) picks up sound and changes it to an electrical signal at mic level (a few millivolts).
2. The mic signals go into a stereo **mic preamp**. This device amplifies the mic-level signals up to line level (about a volt). It is built into a Gemini iKey Plus.
3. The stereo line-level signals go into an **audio interface** (such as the Gemini iKey). It includes an A/D (analog-to-digital) converter that

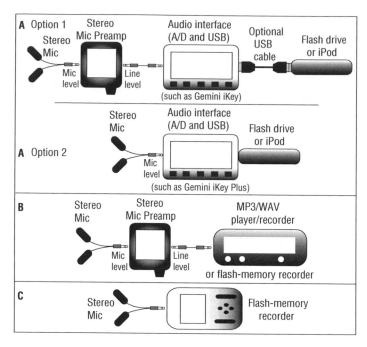

Figure 1-6 Three equivalent flash-memory recording systems: (a) stereo mic, mic preamp, interface, and flash drive; (b) stereo mic, mic preamp, and MP3/WAV player/recorder; and (c) stereo mic and flash-memory recorder.

changes the analog line-level signal into a digital signal (a series of ones and zeros).

4. Also in the interface is a USB encoder, which converts the digital audio signal into USB format. USB is a connection for high-speed transfer of digital data from one device to another. The USB signal data comes out of the USB connector in the interface.

5. The USB cable connects to a USB port in a **flash drive (USB flash memory)**, which records the audio in flash memory as an MP3 or WAV file. An alternative to a flash drive is an **MP3 player** that can record MP3 or WAV files. Some newer iPods can record stereo WAV files at 44.1 kHz in extended mode..

In summary, the signal flow in Figure 1-6(a) is mics > mic preamp > interface (A/D converter and USB encoder) > flash drive with flash memory. During playback, the digital audio plays back from flash memory, is converted to analog audio by a D/A (digital-to-analog) converter, and comes out the headphone jack.

Let's take a closer look at the audio interface in Figure 1-6(a). The battery-powered Gemini iKey USB audio interface (about $100) has a line input and a USB output (www.ikey-audio.com). Its noise floor is not up to professional standards, but it is quiet enough for hobbyist use. The iKey Plus ($169) has a MIC input.

Some of the components in the flash-memory recording system have been combined into one device. For example, Figure 1-6(b) shows a stereo mic plugged into a mic preamp, which feeds the line input of a portable MP3 player that can record MP3 or WAV files. One example of an MP3 player/recorder is the Cowon Systems iAudio U2 (www. eng.iaudio.com, about $130), which includes USB 2.0 and records on a flash-memory card up to 2 GB. Other units are the iriver T30-1GB for $119 (www.iriver.com) and the NOMAD Jukebox 3 for $259 (www. nomadworld.com).

As shown in Figure 1-6(b), you also need a stereo mic and battery-powered mic preamp to record live music with an MP3/WAV player/recorder. The mic preamp amplifies the mic signal and sends it to the player's line-in connector. One mic preamp is made by Archos (www. hotmp3gear.com/Microphone.htm). Selling for $49, the Archos preamp is not professional quality, but it might be all you need. A step up in quality is the Church Audio ST20A preamp for $80 (www.church-audio.ca or http://stores.ebay.com/church-audio). The SP-preamp-4 by The Sound Professionals is a pro-quality unit for $199 (www.soundprofessionals.com). So is the PA-24NJ preamp from www.sonicstudios.com.

Check out these specs when shopping for a mic preamp:

- *S/N ratio*: 90 dB is fair; 110 dB or higher is excellent.
- *Frequency response*: 20 Hz to 20 kHz ±2 dB is fair; ±1 dB or less is excellent.
- *THD*: 0.5% is fair; 0.06% or less is excellent.

If no specs are given, the unit probably is not professional quality.

The components in a flash-memory recording system have been combined in other ways. For example, Figure 1-6(c) shows a stereo mic plugged into a flash-memory recorder, which we discussed earlier. Built into the recorder are mic preamp, A/D converter, and flash memory. Audio is recorded (stored in memory) as an MP3 or WAV file. The signal flow is mics > mic preamp > A/D converter > flash memory. A USB port in the recorder can be used to transfer the audio recordings (files) to a computer for editing.

When assembling a flash-memory recording system to record live music, consider these options:

- If you already. have a flash drive, iPod, or MP3 player/recorder with a USB port, add a stereo mic, mic preamp, and audio interface. Omit the MIC preamp if using iKey Plus.
- If you already have a portable MP3/WAV recorder with line-in, add a stereo mic and mic preamp.
- If you already have a PDA (Personal Digital Assistant), add a stereo mic, mic preamp/A-D converter, PDAudio card, and PDAudio recording software (all available at www.core-sound.com).
- If you want the convenience of an all-in-one system, use a stereo mic and a flash-memory recorder. Some recorders come with a stereo mic built in. Dedicated flash-memory recorders offer higher sound quality than the multi-component systems.
- If you are recording directly off a mixing board, you can omit the stereo mic and mic preamp in any of the systems above. Simply plug the board's tape-out or rec-out connectors into the line-in connector(s) of your recording device, using an appropriate adapter cable.

One source of stereo mics, mic preamps, interfaces, and recorders is The Sound Professionals (www.thesoundprofessionals.com). They offer a wide variety of devices at different price/quality levels.

MiniDisc Recorder

Now that we covered flash-memory recording systems, let's look at other stereo recording devices.

A Hi-MD MiniDisc recorder lets you record uncompressed CD-quality wave files on low-cost Hi MD MiniDiscs. You get up to 94 minutes recording time on a 1 GB disc. Examples include the MZ-M10 ($300) and MZ-M200 ($400). They come with a stereo microphone and earbud headphones. Because MiniDiscs cost little and are removable, a MiniDisc recorder is a good choice if you're recording in the field for a long time and can't dump a flash-memory recording to a computer. You must use the provided Sony software to copy MiniDisc files to a computer. MiniDisc recorders can skip if bumped, so you need to hold the recorder steady.

DAT Recorder

The DAT recorder, which records high-quality digital audio on a small DAT tape cassette, is becoming obsolete. Sometimes you can find DAT recorders on eBay at very low prices.

Laptop, Recording Software, and Audio Interface

Another stereo recorder is a laptop computer with recording software (Figure 1-7). To get audio into the computer, use a two-channel *audio interface*. This is a mic preamp with two mic inputs and a USB or FireWire port, which connects to a similar port in your laptop. If your computer lacks that port, get a USB or FireWire PC Card adapter. It is a PCMCIA card with a USB or FireWire port. Plug the card into your laptop and connect its port to the audio interface. Another option is a CardBus card, which is an advanced PCMCIA card with faster speed.

Some examples of a two-channel USB audio interface are the M-Audio Transit and Fast Track Pro (Figure 1-8) (www.m-audio.com).

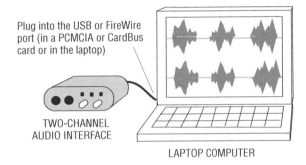

Plug into the USB or FireWire port (in a PCMCIA or CardBus card or in the laptop)

TWO-CHANNEL
AUDIO INTERFACE

LAPTOP COMPUTER

Figure 1-7 An audio interface plugged into a laptop computer via a USB connection.

Figure 1-8 M-Audio Fast Track Pro, an example of a two-channel audio interface (courtesy: M-Audio).

15

USB or FireWire PC Card adapters and CardBus adapters can be found in a Google search (www.google.com). Recording software is described under the heading "Computer DAW Recording Systems" later in this chapter.

When a laptop recording is done, you are ready to edit it. You don't have to transfer the wave files from recorder to computer as you do with other methods.

Headphones or Earphones

Headphones or earphones let you know whether the mics are working correctly, and let you hear what the mics are picking up. Room noises that you wouldn't otherwise notice become obvious when you listen on headphones. Also listen for buzzes, distortion, and crackles from bad cables or connections. If the band and PA are loud, it is hard to hear what's being recorded unless you use *isolating headphones* (Remote Audio HN-7506) or *isolating earphones* (Etymotic ER-4S, ER-4P, and ER-6i; Shure E3 and E4.)

Multitrack Recording Systems

Now we get into professional multitrack techniques, which can offer better sound than stereo techniques. Figure 1-9 shows the parts of a typical multitrack recording system. Several mics are used, *each close to an instrument or singer*. This is the signal flow:

1. Musical instruments produce sound.
2. Microphones pick up the sound and change it into electrical signals. Because each mic is close to its instrument, it picks up very little background noise and room reverberation.
3. Mic choice and placement affect the tone quality and the amount of leakage that are picked up. (*Leakage* is unwanted sound from instruments other than the one the mic is aimed at.)
4. Mic cables carry the mic signals.
5. The mic cables plug into a *stage box*: a box with multiple mic connectors. Wired to the connectors is a long multiconductor cable called a *snake*.
6. The snake connectors plug into a mixing console. If the musical event is reproduced over a sound system, the mixing console is the one

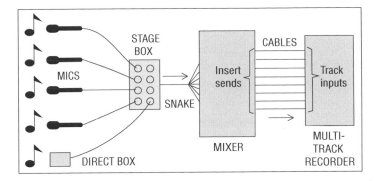

Figure 1-9 Signal flow in a typical multitrack recording system.

used for sound reinforcement. It amplifies each mic signal up to a higher voltage called *line level*.

7. The amplified signal of each mic appears at an *insert-send* connector on the PA mixer.

8. Cables plugged into the insert-send connectors carry the signal to a multitrack HD recorder.

9. The multitrack unit records each mic's signal on a different track. You mix these tracks later back in your studio.

An alternative method is to connect the mic cables to a splitter, which sends each mic's signal to two mixers: one for PA and the other for recording. We cover this equipment later in this chapter.

Equipment for Multitrack Recording

Let's explore each piece of gear from left to right in Figure 1-9.

Microphones

Mics for multitrack recording have an XLR connector and a low-impedance-balanced output. *Condenser* mics sound great on cymbals, acoustic instruments, and vocals. *Dynamic* mics with a "presence peak," a rise in the frequency response around 5 kHz, are a popular choice for guitar amps, drums, and vocals. *Ribbon* mics sound good on vocals, horns, and guitar amps. Of course, you can use any mic on any instrument if it sounds good to you. Just keep ribbon mics out of kick drums because the ribbon is fragile.

Omni mics are seldom used for miking pop groups because omnis pick up too much feedback from the PA system and too much leakage (unless they are placed very close to the sound source). Unidirectional mics (cardioid, supercardioid, and hypercardioid) pick up less feedback and leakage.

If you are recording a band that has a high-quality sound system, typically the sound company's mics supply the signals for your recording.

A *direct box* (DI) can replace a microphone on electric instruments. It has a phone jack (socket) input and a male XLR output. You connect the input to an electric bass, synthesizer, or instrument pickup; and connect the output to a mic connector on the stage box.

Two handy accessories reduce mic noises: Foam *pop filters* or windscreens on vocal mics keep "P-pops" under control, while *shock mounts* reduce pickup of mic-stand thumps and floor thumps. *Clamp-on mic mounts* attach to drums to reduce clutter. They are made by Mic-Eze (www.accetera.com) and others.

Stage Box and Snake

You could run several mic cables from each mic to the PA mixer, but that is unnecessary. Plug the mic cables into a *stage box*. This is a metal box or chassis with several female XLR connectors (Figure 1-10). Each connector

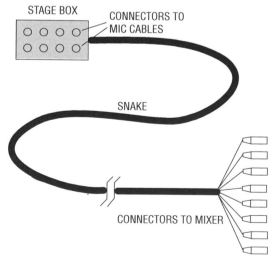

Figure 1-10 A stage box and snake.

has a number. The box is wired to a single multiconductor cable called a *snake*. At the far end of the snake, the cable divides into several individual cables, each with a corresponding numbered male XLR-type connector. These male XLR's plug into the PA mixer mic inputs.

Mixer

A mixer or mixing console (board or desk) is an audio control panel. Each mic connector from the snake plugs into its own preamp built into the mixer. These preamps amplify the signals of all the microphones up to line level so they can feed the line inputs of a multitrack recorder.

Here are the main controls in each mixer input module (mic channel) (see Figure 1-11):

- *Input trim or gain*: Used to adjust the mic preamp gain (amplification). This affects your recording level on each track.
- *Fader*: A sliding volume control for each microphone. This affects the listening level but not the recording level. If you have a separate recording mixer, you can use its faders to set up a monitor mix over headphones.
- *EQ or equalization*: Control of the bass, midrange, and treble. Low-frequency EQ controls the bass (roughly 20–150 Hz); mid-frequency

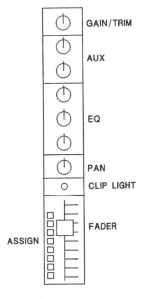

Figure 1-11 Typical input module in a mixing console.

19

EQ controls the midbass (150–500 Hz) through the upper midrange (2–5 kHz); and high-frequency EQ controls the treble (5–20 kHz). These controls do not affect the recording unless the signals sent to the recorder are post-EQ.

- *Pan*: If you have a separate recording mixer, its pan pots let you place the instruments left, center, and right in the stereo stage. Then you can distinguish their sounds more easily.

- *Aux (Auxiliary)*: In the PA console, the aux knobs are used either to control the amount of effect (reverb, echo, chorus, and so on) for each mic, or to set up monitor mixes. The aux controls don't affect the recording.

- *Assign*: Lets you route or send each mic signal to the desired output channel (bus). For example, you might want to send all the drum mics to two buses that feed two recorder tracks. Those tracks would record the stereo drum mix.

- *Insert connectors*: These connect to the multitrack recorder inputs. Each channel's insert-send connector supplies an amplified line-level signal from the mic plugged into that channel. Sometimes the PA engineer plugs a compressor into an insert connector to automatically control the volume of a microphone.

Multitrack HD Recorder

This device accepts the signals from the mixer insert connectors and records each signal on a separate track. The unit records up to 24 tracks on a built-in HD (Figure 1-12). Recorders can be linked to get more recording tracks. You can mix the tracks with an external mixer or copy the track recordings to a computer for editing and mixing. Some examples are Alesis ADAT HD24XR, Tascam MX-2424, Otari DR-100, iZ Technology RADAR V, Fostex D-2424LV, and Mackie HDR 24/96.

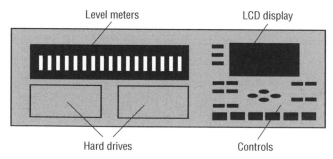

Figure 1-12 A multitrack HD recorder.

Instead of the mixer and HD recorder, you might use one of these systems instead (Figure 1-13):

- a recorder-mixer (a mixer with a built-in HD recorder);
- a mixer feeding an audio interface, connected by USB or FireWire to a laptop;

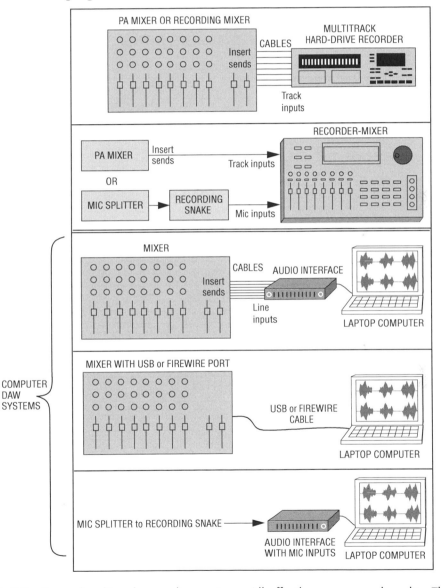

Figure 1-13 Several multitrack recording systems. All offer the same sound quality. The snake connectors plug into any one of these systems.

21

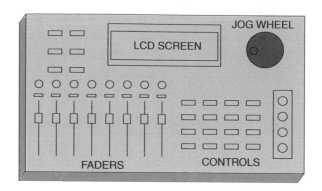

Figure 1-14 A recorder-mixer.

- a mixer with a FireWire or USB port connected to a laptop;
- an audio interface with mic preamps, connected by USB or FireWire to a laptop.

We will examine each option.

Recorder-Mixer Option

Shown in Figure 1-14, a recorder-mixer is an alternative to an HD recorder. Also called a stand-alone digital audio workstation (DAW), portable studio, or digital multitracker; it combines a mixer and multitrack HD recorder in a single package. Some units can record and play up to 36 tracks. The mixer includes faders (volume controls) for mixing, EQ or tone controls, and aux sends for effects (such as reverb). An LCD screen displays recording levels, waveforms for editing, and other functions. For on-location work, you might need to record on all tracks at once, so make sure the recorder-mixer can do that. Some manufacturers of HD recorder-mixers are Roland, Korg, Fostex, Boss, Yamaha, Zoom, Akai, and Tascam.

Computer DAW Recording Systems

Another multitrack recording system is a computer DAW. A computer with recording software can record multiple tracks of audio. For on-location work, a laptop computer is easily portable. Or you might use a desktop computer in a shock-mounted rack. (*Caution*: A computer is more likely to crash than an HD recorder or recorder-mixer.)

PC Audiolabs (www.pcaudiolabs.com) and Sweetwater Sound (www.sweetwater.com) offer custom and preconfigured DAW systems.

To record 24 or more tracks at once, the computer must be fast. At a minimum it should have a 1.2 GHz CPU, 512 MB RAM, an operating system that provides 32-bit or higher data paths, and a USB 2.0 or FireWire port. That port can be in the computer, or it can be part of a CardBus FireWire or USB 2.0 adapter. Another recommendation is an external FireWire HD for the audio data. Ideally that HD has a rotational spindle speed of 7200 rpm or greater and an internal buffer of at least 8 MB. When you are done recording, you can plug that HD into your studio DAW for editing and mixing.

According to tests at Sweetwater Labs, a Mac mini or PowerMac dual G5 can record at least eighty 48 kHz/24-bit tracks simultaneously via FireWire to an internal SATA HD.

A slow computer might cause clicks or dropouts (silent spots) in the audio, and the recording might stop. If you can't record enough tracks at the same time without dropouts or clicks:

- Increase the I/O buffer size in your recording software.
- Reduce video acceleration: in a PC, right-click the desktop > Properties > Settings > Advanced > Troubleshoot. Reduce Hardware Acceleration as much as you can without degrading the display of your audio program.
- Defragment your audio HD.
- Record at 44.1 kHz instead of higher rates.
- As a last option, record at 16-bit resolution instead of 24 bits.
- Check out www.tweakXP.com, www.Tweak3D.net, and www.extremetech.com for computer tweaks that increase speed.

Some examples of recording software are Adobe Audition, MOTU Digital Performer (Mac only), Steinberg Cubase SX and Nuendo, Digidesign Pro Tools, Apple GarageBand, Cakewalk Music Creator, Home Studio and Sonar; Emagic Logic (for Mac only), Sony Pictures Digital Vegas and Sound Forge, BIAS Deck, Pro Tracks Plus, Magix Samplitude, Mackie Tracktion, Magix Sequoia, RML Labs SAW Studio, and Audacity (which is freeware).

Several types of DAWs are explained below.

DAW Option 1: Mixer, Interface, and Laptop

In this system, the mixer feeds a *multichannel audio interface* connected to a computer with audio recording software (Figure 1-15). The PA mixer or

23

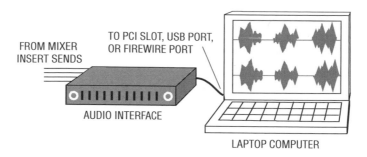

FROM MIXER
INSERT SENDS

TO PCI SLOT, USB PORT,
OR FIREWIRE PORT

AUDIO INTERFACE

LAPTOP COMPUTER

Figure 1-15 Mixer to interface to laptop.

recording mixer amplifies each mic signal up to a line-level signal, and those line-level signals plug into the line inputs in the multichannel audio interface. The interface gets audio into and out of the computer. It converts the analog signals from the mixer to digital and feeds that digital signal to the computer's USB/FireWire port or USB/FireWire CardBus adapter.

Some models of multichannel interfaces are the M-Audio FW1814, MOTU Traveler, MOTU UltraLite, Echo AudioFire8/AudioFire 12, RME Multiface II, and Edirol UA-101/FA-101. Other manufacturers are Korg, Maya, Apogee, and Digidesign. USB or FireWire CardBus adapters can be found in a Google search. RME's FireWire CardBus connects only to RME's interface.

Interfaces typically have eight channels. You can add more interfaces to record more tracks. A typical connection is Interface 1 > FireWire > Interface 2 > FireWire > Laptop FireWire port.

DAW Option 2: USB or FireWire Mixer and a Laptop

A few mixers have a FireWire or USB port that connects to a computer for multitrack recording (Figure 1-13, middle). The signal of each input channel on the mixer (plus the stereo mix) is sent to your computer, and a stereo output returns from the computer for monitoring. It's like having a mixer and a multichannel interface in one compact chassis. Examples are Mackie Onyx Series, Alesis MultiMix Series, Phonic Helix Board 18 FireWire, and Yamaha MW USB Series Mixing Studios.

DAW Option 3: Interface with Mic Inputs and a Laptop

In a setup using a mic splitter (explained next), you can use any one of the recording systems described before. Another option is to use one or more

24

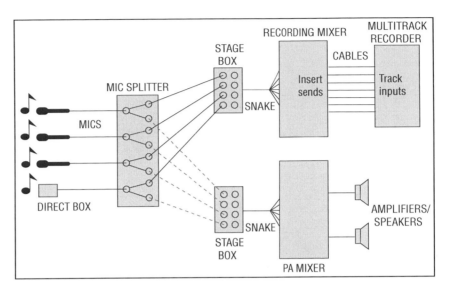

Figure 1-16 Splitting the mic signals to the recording mixer and PA mixer.

audio interfaces with eight mic preamps built in (Figure 1-13, bottom). Plug the recording snake into the interface mic inputs and connect the interface to a computer. Some products with eight mic inputs are the MOTU 896HD, Presonus Firepod, and Alesis io|26. You might prefer to use several mic preamps (maybe in one chassis) feeding an audio interface or an HD recorder.

Mic Splitter

Another way to make multitrack recordings is with a mic splitter and a recording mixer. You connect the mic cables to a multichannel splitter, which sends each mic's signal to two destinations: the recording mixer's snake and the PA mixer's snake (Figure 1-16). This gives you independent control of each microphone. The splitter has one XLR input and two or three XLR outputs per mic, sort of like a Y-cord.

Inside the splitter, the input XLR is wired to a 1:1 transformer (Figure 1-17). The splitter has three feeds or outputs: one direct and two isolated. Wired directly to the mic input connector, the direct XLR output connects to the mixer that supplies phantom power.

The two transformer-isolated XLR outputs connect to the other mixers. Since the transformer electrically isolates the three mixers, phantom power from one mixer can't get into the other mixers. Neither can any

25

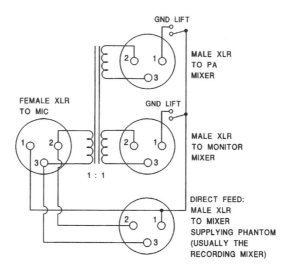

Figure 1-17 Transformer-isolated microphone splitter.

radio-frequency interference (RFI). Ground-lift switches in the splitter are used to prevent "ground loops" and their resulting hum.

Good transformer-isolated mic splitters typically cost $30–$80 per channel. Examples are ProCo MS2, Rolls MS20, and Whirlwind SP1X2. For elaborate productions, multichannel mic splitters (such as the Whirl-wind SB series) offer up to 30 inputs in a single stage box.

Because high-quality mic splitters are expensive, you might want to rent them. Some PA consoles have mic splitters built in. Low-cost transformer splitters are not worth purchasing because their small transformers have poor bass response.

To save money some people use Y-splitters (Figure 1-18) instead of transformer-isolated splitters. The resistors shown in Figure 1-18 prevent mic loading and improve isolation, while the ground lifts prevent ground loops and block phantom power. Unlike a transformer splitter, a Y-splitter does not block RFI.

If you employ the mic splitter approach, you can use any of the recording systems described before. For example, you could plug the recording snake into:

- a recorder-mixer,
- a mixer feeding an HD recorder,
- a USB or FireWire mixer feeding a laptop,
- an interface with eight mic preamps feeding a laptop.

26

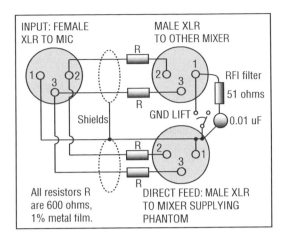

INPUT: FEMALE
XLR TO MIC

MALE XLR
TO OTHER MIXER

RFI filter

51 ohms

0.01 uF

GND LIFT

Shields

All resistors R
are 600 ohms,
1% metal film.

DIRECT FEED: MALE XLR
TO MIXER SUPPLYING
PHANTOM

Figure 1-18 A resistor-isolated Y-cable (Y-splitter or hard-wired splitter).

Headphones, Earphones, or Speakers

Headphones, earphones, or monitor speakers in an isolated room let you check the signal quality. Listen for hum, crackles, hiss, RFI, and distortion. See the heading "Headphones/Earphones" in the section "Equipment for Stereo Recording."

Purchasing Equipment

Some Web stores in the US that sell audio gear are bswusa.com, sweetwater.com, musiciansfriend.com, americanmusicalsupply.com, wwbw.com, guitarcenter.com, samash.com, samedaymusic.com, and zzounds.com. Some United Kingdom (UK) sites are proaudiostore.co.uk, dv247.com, playrecord.net, and bonnersmusic.co.uk. If you search for the product model on www.froogle.com, you get a list of vendors and prices. Audio and music stores in your locale are another option.

So far we have covered the equipment for stereo and multitrack recording. Next we explore recording techniques in Chapter 2.

2

RECORDING TECHNIQUES FROM SIMPLE TO COMPLEX

You can record pop-music concerts in several ways. Listed below are a range of techniques from simple to elaborate. In general, sound quality improves as the recording setup becomes more complex:

- Record off the board (PA mixer) into a portable stereo recorder.

- Record with two mics into a portable stereo recorder.

- Record with a four-tracker: record with a stereo mic on tracks 1 and 2, while recording the PA mixer output on tracks 3 and 4. Mix the tracks later.

- Feed the PA mixer's insert-send signals to a multitrack recorder. Edit and mix the recording back in your studio.

- Use a mic splitter on stage to feed the PA mixer, monitor mixer, and recording mixer. Record to multitrack. Edit and mix the recording back in your studio.

- Do the multitrack recording in a van or truck.

We'll describe each method and list its pros, cons, and equipment. Then you can decide what's best for you and try it out.

Record Off the Board

Let's start with the easiest technique: connect the PA mixing console (board) to a stereo recorder. Some people record performances off the board at gigs, festivals, or concerts, and—let's hope, with the artists' permission—they put the recordings on the Web as MP3 files.

Using cables with the appropriate mating connectors, connect the PA mixer's tape-out or two-track-out connectors to your stereo recorder's line input(s). If your recorder's line input is a stereo mini phone jack, use an adapter cable: either two 1/4-inch phone plugs to a stereo mini phone or two RCA connectors to a stereo mini phone. They are available at Radio Shack and other electronic suppliers. Outside the US, if your recorder's line input is a stereo mini socket, use an adapter cable: either two 6.35-mm jack plugs to a stereo mini jack, or two phono connectors to a stereo mini jack.

If your recording device has balanced line-level inputs (female XLR or TRS, tip/ring/sleeve), connect those inputs to some spare master outputs on the PA mixer. Another place to connect is the insert-send connectors of the master output channels.

Caution: If you plug into an insert-send connector, you will kill the PA mixer's output signal unless you plug in halfway, to the first click. If that doesn't work, get two Y-cables with phone plugs (called "jack plugs" outside the US). At each mixer output channel, insert the Y-cable fully into the insert-send connector. Connect one leg of the Y to the insert-return connector. Connect the other leg of the Y to your recording device line input.

Pros: Simple, cheap, and fast.

Cons: Quality depends on the PA company's mics and board. The mix depends on the mixer operator's skill—you take what he or she sends you.

Equipment: Mixer-to-recorder cables, stereo recorder, and headphones. Your stereo recorder may be a flash-memory recorder, MP3 player/recorder, stand-alone CD burner, cassette deck, computer and audio interface, digital audiotape (DAT) recorder, or MiniDisc recorder.

The recorded mix might be poor with this method. Here's why: the sound mixer hears a combination of the band's instruments, the stage monitors, and the house speakers. So the board mix is intended to *augment* the sound of the instruments and monitors on stage—not to sound good by itself. For example, if the bass-guitar amp is very loud on stage, it will be turned down in the mix that you record off the board. Vocals will be turned up high in the mix because they can't be heard otherwise. Board mixes can sound good if there is not much sound coming off the stage (as with acoustic groups), and the venue is large or outdoors. It helps to monitor the board mix with headphones or isolating earphones to hear what you're recording.

Record with Two Mics

Here's another simple method. Connect two mics to a portable stereo recorder (Figure 2-1). Place the mics in front of the group, set the recording level, and hit Record. You'll capture how the band sounds to an audience.

Pros: Simple, fast, and cheap. If you place the mics a few feet from a folk group or jazz group without a PA system, the sound can be quite good.

Cons: When you record a group with a PA system indoors, the sound will be distant and muddy compared to using a mic on each instrument and vocal. You'll pick up the sound of the PA speakers as well as the band itself. Also, the balance you get depends on the skill of the PA mixer operator.

Equipment: Stereo recorder, headphones or earphones, mic cables, and mics. Four microphone options are listed below:

- A stereo mic on a mic stand.
- Two stand-mounted mics of the same model number (Chapter 8 shows how to arrange a pair of mics to record in stereo).
- A miniature stereo mic that plugs into the recorder, or is built in.
- Two headworn mini mics. These can be clipped to eyeglass frames, either in the ear or on the temples. They make a binaural recording that sounds very realistic when heard over headphones. You could clip the mics to your shoulders instead. Mic models of all these types are described in Chapter 12.

Before going on the road, install fresh batteries and clean the connectors with isopropyl alcohol or DeOxit from Caig Labs (www.caig.com). Connect all your equipment and make a trial recording to make sure everything works. Approximate recording levels can be set in advance by recording loud music from your home stereo speakers or studio monitors. You can then place the stereo recording system in a backpack, suitcase, or cloth

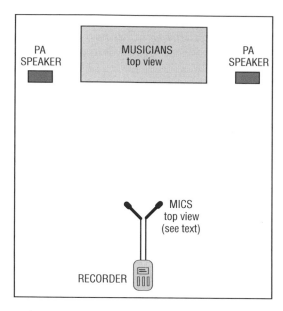

Figure 2-1 Recording a pop group with two microphones.

bag. You might even stow your gear in a hunter's vest or photographer's vest with pockets.

If possible, record in a room where the audience is attentive and the background noise is low. You might visit some potential venues in advance to check out the noise and acoustics. Avoid very live rooms because they can make the recording muddy.

To prevent crackles or loss of audio, strain-relieve the mic cable. Use tape or Velcro to fasten the mic cable to the recorder, so that the cable doesn't get pulled out accidentally. Check that the mic connector is plugged in all the way.

If your recorder can record MP3 or WAV files, consider these options. Unlike MP3 files, WAV files are uncompressed, linear PCM recordings. WAV recordings sound better than MP3 recordings, but WAV files consume a lot of memory: about 10 MB/minute for a 16-bit/44.1-kHz stereo recording. If you select WAV, set the word length or resolution: 16-bit is CD quality and 24-bit is higher quality. Also set the sampling rate: 44.1 kHz is CD quality (good enough for professional recordings), 96 kHz is higher quality, and 192 kHz is state-of-the-art quality.

MP3 files typically consume about 1/11 to 1/4 as much memory as WAV files, depending on the bitrate setting (in kilobits per second or kbps). A setting of 128 kbps results in good audio quality (cassette quality), 192 kbps is very good (near-CD quality), and 256–320 kbps is excellent (CD quality); 64 kbps mono is the same audio quality as 128 kbps stereo; 64 kbps stereo WMA (Windows Media Audio) is the same quality as 128 kbps stereo MP3.

Listed below are approximate recording times on a 1-GB flash-memory card (for a 930-MB file size). Double these times for a 2-GB card.

24-bit/44.1-kHz stereo WAV	1.0 hour
16-bit/44.1-kHz stereo WAV	1.5 hours
256 kbps stereo MP3	9.0 hours
128 kbps stereo MP3	16.5 hours

Be sure you have enough free space on your flash-memory card before going on location.

When you monitor the mics with headphones or earphones, you'll hear the room acoustics and any background noise (audience, air conditioning, traffic). The closer the mics are to the group, the clearer and cleaner the sound will be. In other words, close placement captures more of the music and less of the room sound and background noise. Try to place the mics as close as possible to the stage where you still pick up the house PA speakers well, about a stage-width away from the stage (Figure 2-1). An alternative placement is near the house mixing console because the balance there is what the sound mixer intended. Keep the mics away from any bars or other noise sources. Some concerts have a tapers' section where you are allowed to place your mics. If you are recording your own band, you could place the recorder and mics on stage (on a stool or mic stand) and record the sound of the monitor speakers.

Another way to cut down on room sound is to use a pair of cardioid or supercardioid mics aimed straight ahead at the musicians and spaced about 2 feet apart (or headworn). This will give a closer, clearer sound than an XY stereo mic or a pair of omni mics.

If there are dancers near the stage and the ceiling is low, you might try boundary mics (such as two Pressure Zone Microphones (PZMs)) gaffer-taped to the ceiling or mini mics hung from the ceiling.

To record a small folk group or acoustic jazz group, place the mics about 3 to 6 feet from the ensemble (Figure 2-2). If the group plays in a circle (as in an outdoor jam), try placing the mics in the center. Or just walk

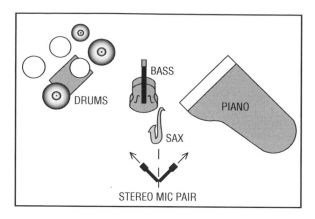

Figure 2-2 A method of stereo miking a jazz group.

around with the mics while monitoring the mics over headphones. Find a spot where you hear a good balance and put the mics there. Ask the musicians' permission to record them during a break in the music.

The recorder might have a record-level switch labeled "manual" and "auto." Set it to "manual" in order to retain the dynamics of the performance. If the switch is labeled "AGC" (Automatic Gain Control), set it to "off."

The recorder also has a meter that shows recording level. Set the level so that the meter reads about −6 dB maximum. That allows some headroom for surprises. Peak levels up to 0 dB are okay, but levels above 0 dB result in distortion. (Some portable recorders include a limiter that prevents recording levels above 0 dB.) After setting the recording level, leave it alone as much as possible. If you must change the level, do so slowly and try to follow the dynamics of the music.

If your recording is distorted even though you did not exceed a 0 dB recording level, either the sound was louder than the mic could handle (not likely), or the mic signal overloaded the mic preamp in the recorder. You can prevent mic-preamp overload with a mic-gain switch or pad (attenuator), available in most recorders. Use low gain for loud sound sources (rock bands); use high gain for quiet sound sources (acoustic musicians). If you need to set the recording-level control less than 1/3 up to achieve a 0 dB recording level, use the low-gain setting or switch in the pad.

If you are using large-diaphragm condenser mics plugged into a phantom power supply, you might need to plug the output of the supply into the recorder line input—rather than the mic input—to prevent distortion.

You might be able to record an acoustic group on location without a PA system or audience. This gives you the freedom to make adjustment to improve the sound. Here is a suggested procedure:

1. Adjust the acoustics around the instruments. If the room is too live (reverberant), put up some packing blankets, comforters, rugs, acoustic foam, or cushions. Often a good-sounding spot for the musicians is near the center of a large room.
2. Place the musicians around the stereo mic pair where you want them to appear in the recording. For example, you might place two singing guitarists on the left and right, with bass in the center.
3. Experiment with microphone height to vary the vocal/guitar balance. Try different miking distances to vary the amount of ambience or room sound. A 3- to 6-foot distance is typical.
4. While the musicians are playing and during playback, monitor the mic signals with headphones. If some instruments or vocalists are too quiet, move them closer to the mics, and vice versa, until the balance sounds right.
5. If someone makes a mistake, either record another take of the entire tune, or record starting from a few bars before the mistake and edit the takes together later.

Figure 2-3 shows a clever recording method using two omnidirectional mics placed to get a good balance of a jazz group. Engineer Gert Palmcrantz used this technique to record the Ludvig Berghe Trio. Details can be found at www.discmakers.com/music/pse/2005/jarvis.asp, and recordings at www.moserobie.com.

Once you copy the recorded files to your computer hard drive, you can edit the recording and adjust its tonal balance (equalize it) with digital audio workstation (DAW) recording software or with Harmonic Balancer software. You might cut 1–6 dB around 300 Hz to reduce boomy reverb and get a clearer recording. Mini stereo mics using cardioid mic capsules tend to sound weak in the bass, but you can boost 3–6 dB at 50–100 Hz to compensate (or whatever amount sounds right). Headworn mics often need a cut around 3 kHz to compensate for the effect of the head. If you have recorded outdoors and the sound is too dry, try adding a little reverb and see if it helps.

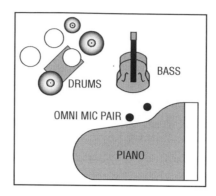

Figure 2-3 A method of stereo miking a jazz group.

A great source of information about two-track recording of concerts and jams is the DAT-heads discussion group (www.solorb.com/dat-heads/). The Steam Powered Preservation Society (www.thespps.org) records folk-music jams and concerts, and posts MP3 files of the recordings along with detailed information on the recording equipment and techniques. Sonic Studios (www.sonicstudios.com/tips.htm) offers a wide array of tips and equipment for this type of recording. Mike Billingsley wrote some fine papers on stereo miking in Preprints 2788 (A-1) and 2791 (A-2), available from www.aes.org.

Record with a Four-Tracker

This method is fairly simple and provides good sound. Put a stereo mic or mic pair at the front-of-house (FOH) PA mixer position. Plug the mic connectors into mic inputs 1 and 2 of a four-track recorder-mixer (or any size recorder-mixer). Connect the PA mixer's tape output or two-track outputs to line inputs 3 and 4 (Figure 2-4). Mix the recording to stereo back in the studio.

> **Pros:** Fairly simple, cheap, and fast. Good sound quality.
>
> **Cons:** You can't control the mix among instruments. Sound quality depends on the PA system and the sound mixer's skill.
>
> **Equipment:** Stereo mic or matched mic pair, stereo bar, mic stand, mic cables, Y-cables, mixer-to-recorder cables, headphones, recorder-mixer (such as models by Tascam, Boss, Zoom, Fostex, or Korg).

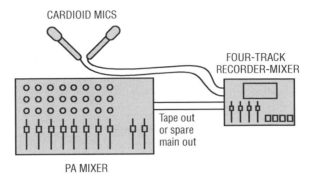

Figure 2-4 Recording two mics and a PA mix on a four-track recorder-mixer.

The FOH mics pick up the band as the audience hears it: lots of room reverb, lots of bass, but rather muddy or distant. The PA mixer output sounds tight and clear, but typically is thin in the bass. Luckily, a mix of all four tracks can sound surprisingly good. Tracks 1 and 2 provide ambience and bass; tracks 3 and 4 provide definition and clarity.

When you mix the four tracks, you might hear an echo because the FOH mics pick up the band with a delay (caused by the sound travel time from stage to mics). To remove the echo, import all the tracks into digital recording software, and delay the PA mixer tracks by sliding them a little to the right. Align the waveforms of the mic tracks and mixer tracks at big peaks.

Connect the PA Mixer Insert Sends to a Multitrack Recorder

Now we get into professional techniques. This is an easy way to record, and it offers very good sound quality with minimal equipment (Figure 2-5). Use the mixer gain trims to set recording levels during the sound check. Edit and mix the recording back in the studio.

Pros: Easy, fast, moderate cost, and sounds great. You don't have to mix while recording—instead, mix and monitor back in your studio.

Cons: Sound quality depends on the PA company's mics and mic preamps. You might have to ask the PA operator to adjust the gain trims during the show to prevent recorder overload.

Equipment: Multitrack recorder and mixer-to-recorder cables (such as a short phone-to-phone snake, called "jack-to-jack" snake outside

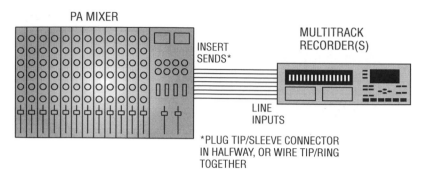

Figure 2-5 Connecting insert sends to a multitrack recorder provides great sound and easy setup.

the US). The multitrack unit can be a hard drive recorder, recorder-mixer, or multichannel audio interface and a laptop.

Connections

This section describes how to connect a multitrack recorder to a mixer's insert-send connectors.

In making a multitrack remote recording, you usually want to record the signal of each mic on a different recorder track. You'll mix those tracks later in the studio.

In the console there are several mic preamplifiers, one per mic, which amplify the mic-level signal up to line level. For each mic channel, this line-level signal typically appears at two connectors on the back of the mixer: *direct out* and *insert send*. That's where to connect to the recorder inputs.

Usually the insert send is the best connector to use. Here's why. Typically the direct-out signal is postfader (Figure 2-6). This means the signal at the direct-out connector comes after the fader, so that the signal is affected by the fader (volume) settings. Any fader movements will show up on your recording, which is undesirable. It's better to connect recorder tracks to insert sends. They are usually prefader, pre-EQ (pre-equalization). So any fader or EQ changes that the PA operator makes will not appear in your recording. However, any changes the PA operator makes in the trim settings during the show will affect your recording levels.

First, find out what kind of insert connectors the PA mixer has, and what kind of input connectors your multitrack recorder has. Buy or make some shielded cables (or a snake) that mate with those connectors. Figure 2-7 shows three ways to wire cables based on the type of insert connector.

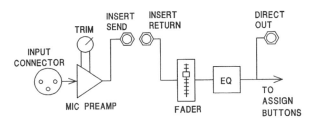

Figure 2-6 Simplified signal flow through part of a mixing console, showing insert and direct out.

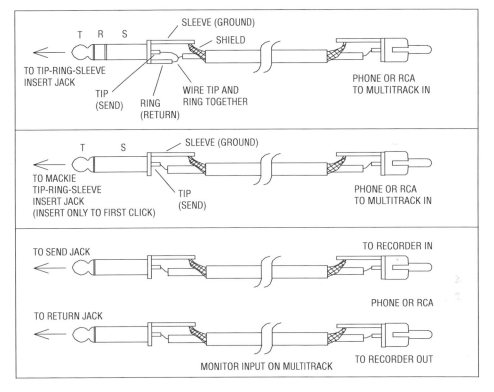

Figure 2-7 Three ways to wire cables based on the type of insert connector.

Some boards have a single TRS insert connector per channel, instead of separate insert-send and insert-return connectors. Usually the tip is send and the ring is return. In the TRS connector that you plug into the TRS insert, wire tip and ring together, and also to the cable hot conductor

(Figure 2-7 top). That way, the insert send goes directly to the insert return. If nothing is connected to the insert return, no mic signal goes through the mixer.

On some PA mixers with a TRS connector, you can use a TS (tip/sleeve) connector (Figure 2-7 middle). Plug it in halfway, to the first click, so you don't break the signal path—the mic signal still goes through the PA mixer. If you plug it in all the way to the second click, the signal does not go through the PA mixer, just to the recorder. Cover the connections with a mixer case cover or board, because someone could bump into the mixer and dislodge a cable.

If the PA mixer has separate send and return insert connectors, connect the send to the recorder track input, and connect the return to the recorder track output (Figure 2-7 bottom). If necessary, set your multitrack recorder to monitor the input analog signal, so that the PA mixer will receive a signal. Another option: use an insert snake with TRS connectors at the mixer end. Carry some TRS-to-dual-TS adapters to handle mixers that have separate connectors for insert send and return.

The insert sends are balanced or unbalanced, and the same is true of the recorder inputs. To connect balanced and unbalanced equipment correctly, see the article "Sound System Interconnection" on the Rane Website, http://www.rane.com/note110.html.

You often encounter PA consoles where some insert sends are tied up with signal processors. You must use those channels' direct-out connectors instead, which usually are postfader (unless they can be switched to prefader). Another option is to "Y" the insert send to your recorder and to the processor input. Or, assign those mic channels to unused groups (buses) and get your recording signals from there.

Some rack-mounted mic preamps have insert connectors. You could plug the mic-snake XLRs into these preamps, then connect the preamp's inserts to the PA mixer line inputs. That way, any gain changes made on the PA console will not affect your recording.

Caution: Any gain changes you make on the mic preamps will affect the PA levels.

If the PA mixer has a FireWire or USB (Universal Serial Bus) port, simply connect the port to a laptop that is running recording software. Set up the software to recognize the mixer as its input/output device. The signal from each mic channel goes to a separate track in the software.

What if you want to record several instruments on one track, such as a drum mix? Assign all the drum mics to one or two output buses in the PA mixer. Plug the bus out insert connector to the recorder track input. Use two buses for stereo. Adjust the faders to get a good drum mix.

Monitoring

To monitor the quality of the signals you're recording, you generally let the PA system serve as your monitor system. But you may want to set up a monitor mix over isolating headphones or earphones so that you can hear more clearly.

Here is a suggested procedure. Connect all the recorder outputs to unused line inputs in the PA mixer, or to a separate mixer. Use those faders to set up a monitor mix. Assign them to an unused bus, and monitor that bus with headphones/earphones. If you can spare only a few inputs, plug in just one track at a time to check its sound quality. Listen closely for any hum, noise, or distortion.

Setting Levels

Set recording levels with the PA mixer's gain-trim or input-attenuator knobs. This affects the levels in the PA mix, so be sure to discuss your trim adjustment in advance with the PA mixer operator. If you turn down an input trim, the PA operator must compensate by turning up that channel's fader and monitor send.

As we said, if the PA operator changes the input trim during the show, these changes will appear in your recording.

Set recording levels before the concert during the sound check (if any!). It is better to set the levels a little too low than too high because during mixdown you can reduce noise but not distortion. A suggested starting level is −10 dB, which allows for surprises. Do not exceed 0 dB because the signal will distort. Also, if you set the recording level conservatively, you are less likely to change the gain trims during the performance. You don't want to hassle the PA operator.

Some recording engineers run each insert-send signal through a potentiometer to set the recording level. By adjusting levels on a rack-mounted panel of potentiometers, you don't have to ask the PA operator to change the gain trims for you. Of course, if you are operating the PA console, you can set the gain trims yourself.

If you have a spare recorder, record a safety copy on it at the same time. This provides a backup in case one recorder fails.

Keep a log as you record, noting the counter times of tunes, level changes, sonic problems, and so on. You can refer to this log when you mix.

Splitting the Mic Signals

Now let's consider a different way to make a multitrack recording. Plug each mic into a mic splitter, which sends the mic signal to two destinations: the PA mixer and recording mixer. This gives you independent control of each microphone. The splitter has one XLR input and two or three XLR outputs per mic. The third output on some splitters goes to a monitor mixer. In Chapter 1 we described transformer-isolated splitters and Y-cable splitters.

As shown in Figure 2-8, connect the outputs from all the splitter channels to the PA snake and to your recording snake. Connect the recording snake to your recording mixer mic inputs. Connect the recording mixer's insert sends to a multitrack recording system of your choice. After the recording is done, you can edit and mix the recording back in your studio.

Pros: Ultimate sound quality. Independent control at each mixer. Consistent sound.

Cons: Complicated. Expensive if transformer splitters are used.

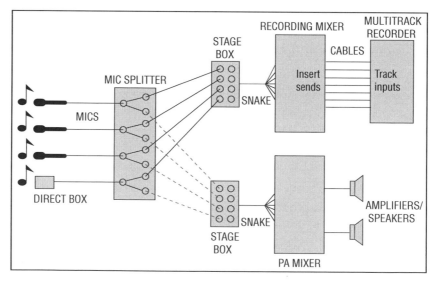

Figure 2-8 Splitting the mic signals to the recording mixer and PA mixer.

Equipment: Mic splitters, maybe mic preamps, mic cables, mic snake, recording mixer, mixer-to-recorder cables, multitrack recorder, headphones, or powered monitors.

Let's explain the advantages of splitting the mic signals. You use your own mic preamps, so you are not dependent on the quality of the PA console mic preamps. Also, you are not hassling the operator about adjusting gain trims. Each mix engineer can work without interfering with the others. The FOH engineer can change trims, level, or EQ and it will have no effect on the signals going to the recording engineer. Another plus: a splitter provides consistent, unprocessed recordings of the mic signals. This consistency makes it easy to edit between different performances. What's more, splitters let you use mic preamps on stage if you wish. That way, the cable from each mic to its preamp is short, which reduces hum and radio-frequency interference.

Using Splitters

To use a splitter or multichannel splitter, plug each mic into a splitter input. Decide which mixer you want to supply phantom power (usually the PA mixer). Connect the splitter's direct outputs to that mixer's snake. Connect one set of isolated outputs to the recording snake, and another set to the monitor snake (if used). Plug the snakes into the mixers. Figure 2-8 shows splitter connections to two mixers.

Splitters have a ground-lift switch on each output channel. This switch connects or disconnects (floats) the cable shield from pin 1 of the XLR connector. When the ground-lift switches are set correctly, you should get no ground loops and their resulting buzzes.

How do you set the ground-lift switches?

1. First, turn *off* phantom power in each console. Turn down all the faders.
2. Make sure the direct feed's ground-lift switches (if any) are set to *ground*, not *lift*. Otherwise phantom power won't work.
3. Go to the mixer connected to the direct feed. Turn on the mixer, switch on phantom power, and bring up each fader to listen for a signal.
4. On the splitter, find the ground-lift switches for the recording-mixer feed. Set them to the position where you monitor the least hum and buzz at the recording mixer.
5. Repeat step 4 for the monitor mixer.

Multitrack Recording in a Truck

Here's the ultimate setup. Each mic signal is split three ways to feed the stage boxes for the recording, reinforcement, and monitor consoles. A long snake is run to a recording truck or van parked outside the concert hall or club. In the truck, the snake connects to a mixing console plugged into a multitrack recorder. The interior of the truck is acoustically treated, and powered Nearfield monitor speakers allow accurate monitoring of the signal. Designing a recording truck is a subject in itself and is beyond the scope of this book.

> **Pros:** The truck contains the console, monitors, and recorders, so you don't need to cart them into the venue. This saves setup time. Also, a truck provides a quiet, controlled monitoring environment.
>
> **Cons:** Complicated and expensive. Requires an AC power connection to the venue circuit-breaker panel or to the house power distro.
>
> **Equipment:** Mic splitters, maybe mic preamps, mic cables, mic snake, recording mixer, mixer-to-recorder cables, multitrack recorder, powered monitors, acoustically treated truck.

There you have a full choice of methods for recording live. Check out all the options, and you'll find a system that works well for your style of recording.

3

BEFORE THE SESSION: PLANNING

Ready to record a live gig? The recording will go a lot more smoothly if you plan what you're going to do. So sit down, grab a pen, and make some lists and diagrams as described here. We'll go over the steps needed to plan a multitrack recording of a concert.

Selecting a Venue

If you can select a place for the recording, use a room where the audience is attentive and enthusiastic, and the background noise is minimal. Visit some potential venues and check out the noise and acoustics. A very live room with lots of reverb can make the recording muddy. Ideally the stage should be large so the performers can spread apart for better separation. A large stage also reduces leakage from wall reflections.

Musical Preparation

Before the recording, band members might want to upgrade guitar strings and drum heads, and repair the "action" and intonation of guitars.

Long jams may be fun to play live, but they can get boring on a CD. Try to keep the songs short and focused. Be sure to check tuning before each song.

The band should consider playing some songs more than once. If a song is played poorly in the first set, the band can re-play it in a later set and re-record it. Put the best performance on the CD. If budget allows, record several shows and pick the best takes for the album or demo. The song sequence can be changed during mastering.

Preproduction Meeting

Call or meet with the PA company and the production company (or just the musicians) who are putting on the event. Find out the date of the event, location, phone numbers, and email addresses of everyone involved, when the job starts, when you can get into the hall, when the second set starts, and other pertinent information. Decide who will provide the mic split, which system will be plugged in first, second, and so on. Draw block diagrams for the audio system and communications (comm) system. Determine who will provide the comm headphones.

If you're using a mic splitter, work out the splitter feeds. The mixer getting the direct side of the split provides phantom power for condenser mics that are not powered on stage. If the house system has been in use for a long time, give its operators the direct side of the split.

Overloud stage monitors can ruin a recording, so work with the sound-reinforcement people and musicians toward a compromise. Explain to the musicians that if they can play with a low monitor level, it will help their recording. In-ear monitors are preferred over floor monitors for live recording because then no monitor sound gets into the mics.

Make copies of the meeting notes for all participants. Don't leave things unresolved. Know who is responsible for supplying what equipment.

Figure 3-1 shows a typical equipment layout worked out at a preproduction meeting. There are three systems in use: sound-reinforcement, recording, and monitor mixing. The mic signals are split three ways to feed these systems.

Site Survey

If possible, visit the recording site in advance and go through the following checklist:

- Check the AC power to make sure the voltage is adequate, the third pin is grounded, and the waveform is clean.

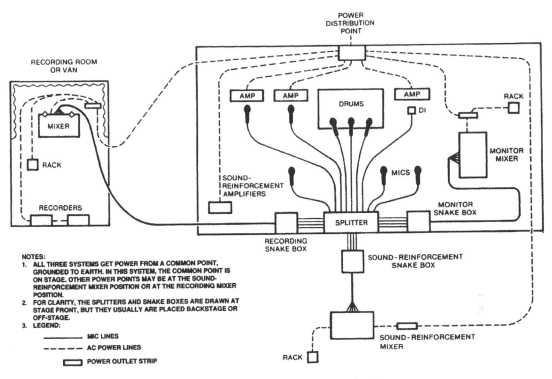

Figure 3-1 Typical layout for an on-location recording of a live concert.

- Listen for ambient noises—ice machines, coolers, 400-Hz generators, heating pipes, air conditioning, nearby discos, etc. Try to have these noise sources under control by the day of the concert.
- Sketch dimensions of all rooms related to the job. Estimate distances for cable runs.
- Turn on the sound-reinforcement system to see if it functions okay by itself (no hum and so on). Turn the lighting on at various levels with the sound system on. Listen for buzzes. Try to correct any problem so that you don't document bad PA sound on your recording.
- Determine locations for any audience or ambience mics. Keep them away from air-conditioning ducts and noisy machinery.
- Plan your cable runs from the stage to the recording mixer.
- If you plan to hang mic cables, feel the supports for vibration. You may need microphone shock mounts. If there's a breeze in the room, plan on taking windscreens.

• Make a file on each recording venue that includes the dimensions and the location of the circuit breakers.

• Determine where the control room will be, if any. Find out what surrounds it—any noisy machinery?

• Visit the site when a crowd is there to see where there may be traffic problems.

Mic List

Now write down all the instruments and vocals in the band. If you want to put several mics on the drum kit, list each drum that you want to mike. As for keyboards, decide whether you want to record off each keyboard's output, or off the keyboard mixer (if any).

Next, write down the mic or direct box you want to use on each instrument; for example:

1	Bass	DI
2	Kick	AKG-D112
3	Snare	Shure Beta 57
4	Hi-hat	Crown CM-700
5	Small-rack tom	57
6	Big-rack tom	57
7	Small-floor tom	Sennheiser MD-421
8	Big-floor tom	421
9	Cymbals overhead left	Shure SM81
10	Cymbals overhead right	SM81
11	Lead guitar	57
12	Rhythm guitar	57
13	Keyboard mixer	DI
14	Lead vocal	Beyer M88
15	Harmony vocal	Crown CM-311A

Make some copies of this mic list. At the gig, place one list by the stage box and the other by each mixer. The list will act as a guide to help you keep things organized.

Track Sheet

Next, decide what will go on each track of your multitrack recorder. If you have enough tracks, your job is easy: just assign each instrument or vocal to its own track. Bass to track 1, kick to track 2, and so on.

What if you have more instruments than tracks? Suppose you have an eight-track recorder, but there are 15 instruments and vocals (including each part of the drum set). You can combine several drum mics into one or two drum tracks (or combine several vocal mics into one or two vocal tracks). That is, you can set up a submix.

Let's say the drum kit includes a snare, kick drum, two rack toms, two floor toms, hi-hat, and cymbals. If you want to mike everything individually, that's nine mics including two for the cymbals. But you don't need to use nine tracks. Assign or group those mics to buses 1 and 2 to create a stereo drum mix. Connect buses 1 and 2 to recorder tracks 1 and 2. At the sound check, you set up a submix of all the drum mics, and assign them to buses 1 and 2. You control the overall level of the drum mix with submaster faders 1 and 2 (also called bus faders or group faders).

Use tracks 3 through 8 for amps and vocals (as in the following example). Feed tracks 3–8 from insert sends.

Track/instrument

1. Drum mix *L*
2. Drum mix *R*
3. Bass
4. Lead guitar
5. Rhythm guitar
6. Keys mix
7. Lead vocal
8. Harmony vocal.

Block Diagram

Now that your track assignments are planned, you can figure out what equipment you'll need. Draw a block diagram of your recording setup from input to output (Figure 3-2). Include mics, mic stands, DI boxes, cables, snake, mixer, multitrack recorder(s), recording media, and powering. On your diagram, label the cable connectors on each end so you'll know what kind of cables to bring. It is a good idea to keep a file of system block diagrams for various recording venues.

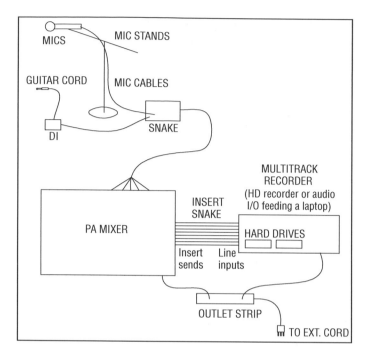

Figure 3-2 Example block diagram of recording setup.

In Figure 3-2, the block diagram shows a typical recording method: feeding PA-console insert connectors to a multitrack hard disk recorder. We'll use this example throughout the rest of the chapter.

Equipment List

From your block diagram, generate a list of recording equipment. Based on Figure 3-2, you would need the following PA and recording gear (not including power amps and speakers):

- Mics
- Mic stands and booms
- Direct box
- Guitar cable
- Snake
- PA mixer

Table 3-1 Storage required for a 1-hour recording.

Number of tracks	Bit depth	Sampling rate	Storage needed
2	16	44.1 kHz	606 MB
2	24	44.1 kHz	909 MB
2	24	96 kHz	1.9 GB
8	16	44.1 kHz	2.4 GB
8	24	44.1 kHz	3.6 GB
8	24	96 kHz	7.7 GB
16	16	44.1 kHz	4.8 GB
16	24	44.1 kHz	7.1 GB
16	24	96 kHz	15.4 GB
24	16	44.1 kHz	7.1 GB
24	24	44.1 kHz	10.7 GB
24	24	96 kHz	23.3 GB

- An insert snake made of eight phone-to-phone cables (called "jack-to-jack" cables outside the US)
- Multitrack recorder
- Outlet strip
- Extension cord
- Hard drives (HDs) (enough capacity for the duration of the gig).

Multitrack audio consumes a lot of disk storage space. Table 3-1 shows the amount of HD or flash-memory space needed for a 1-hour recording with various recording formats.

Don't forget the incidentals: a pen, notebook, flashlight, guitar picks, heavy-duty guitar cords, drum keys, mic pop filters, gaffers tape, console tape, tuner, ear plugs, audio-connector adapters, audio ground-lift adapters, in-line pads, in-line polarity reversers, spare cables, gooseneck lights for the console, spare batteries, bottled water, and aspirin.

Let's explain some of those terms. A ground-lift adapter disconnects the shield at one end of a line-level balanced cable, preventing hum from a ground loop (multiple connections to ground). A pad or attenuator reduces the signal level to prevent distortion from overly hot signals. A polarity reverser reverses the connections to XLR pins 2 and 3 to correct for mics that are wired in opposite polarity.

Instead of duct tape, use gaffers tape, which does not leave a sticky residue. Gaffers tape and console tape are available at www.markertek. com. Also use light-colored electrical tape and a Sharpie pen to identify your mic stands and mic cables.

Bring a tool kit with screwdrivers, pliers, soldering iron and solder, AC-outlet checkers, fuses, a pocket radio to listen for interference, ferrite beads of various sizes for RFI (radio-frequency interference) suppression, canned air to shoot out dirt, cotton swabs and pipe cleaners, and De-Oxit from Caig Labs to remove oxide from connectors.

Check off each item on the list as you pack it. After the gig, you can check the list to see whether you've reclaimed all your gear.

Preparing for Easier Setup

You want to make your setup as fast and easy as possible. Here are some tips to help this process.

Put It on Wheels

Mount your console and recorders in protective carrying cases. Install casters or swivel wheels under racks and carrying cases so you can roll them in. Rolling is so much easier than lifting and carrying.

You might permanently install the HD recorder in an SKB carrying case, which acts as a rack (www.gigcases.com). When a remote job comes up, just grab it and go.

A very helpful item is a dolly or wheeled cart to transport heavy equipment into the venue. Consider getting some lightweight tubular carts. Collapsible carts will store easily in your car or truck. One maker of equipment carts is Rock n Roller, which advertises in the *Musician's Friend* catalog, www.musiciansfriend.com. Another cart is the Remin Kart-a-Bag, www.kart-a-bag.com. Also see www.gigcases.com.

Pack mics, headphones, and other small pieces in trunks or milk crates. You might want to build a mic container: a big box full of foam rubber with cutouts for all the mics. Or construct a wheeled cabinet with drawers for mics, DI's, and speaker cables.

In an article in the September/October 1985 issue of *db* magazine, remote recording engineer Ron Streicher offers these suggestions:

> Especially for international travel, make sure your documentation is up to date and matches the equipment you're carrying. Make a list of everything you take: all the details, such as each pencil, connector, etc.

> Also make sure your insurance is up to date. You need insurance for en route as well as at the destination.

I organize my cases so I know where every item is. They're ready to go anytime and make setup much faster. The cables are packed with their associated equipment, not in a cables case. I check everything coming and going, and try to have 100 percent redundancy, such as a small mixer to substitute for the large console.

Mic Mounts

If you'll be recording a singer/guitarist, take a short mic mount that clamps onto the singer's mic stand. Put the guitar mic in the mount. Also bring some short mounts to clamp onto drum rims and guitar amps. By using these mounts, you eliminate the weight and clutter of several mic stands. Some examples of short mounts are the Mic-Eze units by Ac-cetera (http://ac-cetera.com). They have standard 5/8-27 threads (3/8 inches outside the US) and mic clamps that either spring shut or are screw-tightened.

Snakes and Cables

You can store mic cables on a cable spool, available in the electrical department of a hardware store. Wrap one mic cable around the spool, plug it into the next cable and wrap it, and so on.

A snake can be wrapped around a garden-hose spool, or coiled inside the bottom of a trunk (see Figure 3-3). Commercial snake reels are made by such companies as Whirlwind (www.whirlwindusa.com), Pro Co Sound (www.procosound.com), and Hannay (www.hannay.com).

Use wire ties to join cables that you normally run together, such as PA sends and returns.

Snake hookup is quicker if the snake has a multipin twist-lock connector or "disconnect" (such as Whirlwind W1 or W2). This connector plugs into a mating connector that divides into several male XLRs. Those XLRs

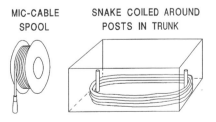

MIC-CABLE SPOOL   SNAKE COILED AROUND POSTS IN TRUNK

Figure 3-3   Some cable-storage methods.

plug into the mixing console. Leave the XLRs in the console carrying case. You'll find that the snake is easier to handle without the XLR pigtails.

For a clean, rapid hookup of drum mics, put a small snake near the drum kit and run it to the main stage box. Snakes are made by companies such as Whirlwind, Pro Co Sound, and Horizon (www.horizonmusic. com).

Check that all your mic cables are wired in the same polarity—pin 2 hot.

You might use three-conductor shielded mic cables (hot, cold, and ground leads in a shield). Connect the shield to ground only at the male XLR end. Also use cables with 100% shielding. Those measures enhance the shielding capability of the shield and reduce pickup of lighting buzzes.

In XLR-type cable connectors, do not connect pin 1 to the shell, or you may get ground loops when the shell contacts a metallic surface.

Label all your line-level cables on both ends according to what they plug into; for example, compressor in, track 12 out, power amp ch.1 in, snake aux1 out. Or you might prefer to number the cables near their connectors. Cover these labels with clear heat-shrink tubing.

Label both ends of each mic cable with the cable length. Put a drop of glue on each connector screw to temporarily lock it in place.

Multitrack Wiring

You can speed the PA console wiring by using a short phone-to-phone snake (jack-to-jack snake outside the US) between the console and your multitrack recorder. When packing, plug the snake into the multitrack recorder and coil the snake inside the recorder's carrying case or rack. In other words, have all your equipment pre-wired. At the gig, pull out the snake and plug it into the console connectors.

Use an insert snake with TRS (tip–ring–sleeve) phone plugs (TRS jack plugs outside the US) at the console end. Carry some TRS-to-dual-TS (tip–sleeve) adapters to handle consoles that have separate connectors for insert send and return. Short snakes for rack and multitrack connections are made by Hosa (www.hosatech.com), Horizon, and Pro Co Sound, among others.

Other Tips

Here are some more helpful hints for successful on-location recordings:

- Plan to use a talk-back mic from the board to the stage monitors during sound checks.

- Hook up and use unfamiliar equipment before going on the road. Don't experiment on the job.

- Consider recording with redundant (double) systems so you have a backup if one fails.

- Walkie-talkies are okay for preshow use, but don't use them during the performance because they cause RFI. Use hard-wired communications headsets. Assistants can relay messages to and from the stage crew while you're mixing.

- Allow for delays at airline security checkpoints.

- Get a public-liability insurance policy to protect yourself against lawsuits.

- Call the venue and ask directions to the load-in door. Make sure that someone will be there at setup time to let you in. Ask the custodian not to lock the circuit-breaker box the day of the recording.

- A few days before the session, check out the parking situation.

- Just before you go, check out all your equipment to make sure it's working.

- Arrive several hours ahead of time for parking and setup. Expect failures—something always goes wrong, something unexpected. Allow 50% more time for troubleshooting than you think you'll need. Have backup plans if equipment fails.

- In general, plan everything in advance so you can relax at the gig and have fun.

By following these suggestions, you should improve your efficiency—and your recordings—at on-location sessions.

Some of the information in this chapter was derived from two workshops presented at the 79th convention of the Audio Engineering Society in October 1985. These workshops were titled "On the Repeal of Murphy's Law—Interfacing Problem Solving, Planning, and General Efficiency On-Location," given by Paul Blakemore, Neil Muncy, and Skip Pizzi; and "Popular Music Recording Techniques," given by Paul Blakemore, Dave Moulton, Neil Muncy, Skip Pizzi, and Curt Wittig.

4

AT THE SESSION: SETUP AND RECORDING

This chapter describes some ways to conduct a multitrack recording session of a concert.

You've arrived at the venue. After parking, offload your gear to a holding area, rather than onstage, because gear on stage will most likely need to be moved.

Learn the names of the PA company crew members and be friendly. These people can be your assets or your enemies. Think before you comment to them. Try to remain in the background and do not interfere with their normal way of doing things (e.g., take the secondary side of the split). A successful remote engineer makes others feel comfortable and exudes confidence.

Power and Grounding Practice

At the job, you need to take special precautions with power distribution, interconnecting multiple sound systems, and electric-guitar grounding.

Power Distribution System

Consider buying, renting, or making your own single-phase power distribution system (distro). It will greatly reduce ground loops and increase

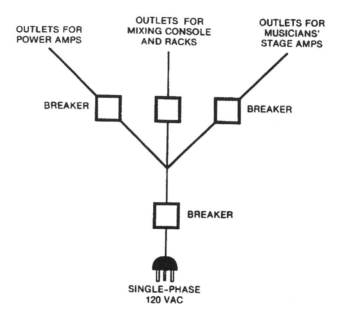

OUTLETS FOR
POWER AMPS

OUTLETS FOR
MIXING CONSOLE
AND RACKS

OUTLETS FOR
MUSICIANS'
STAGE AMPS

BREAKER

BREAKER

BREAKER

SINGLE-PHASE
120 VAC

Figure 4-1 An AC power distribution system for a touring sound system.

reliability. Figure 4-1 shows a suggested AC power distribution system. The amp rating of the distro's main breaker box should exceed the current drain of all the equipment that will be plugged into the distro system. Ask the house electrician to hook up the distro's power cable to the breaker box.

Furman makes a model ACD-100 AC Power Distro. It distributes a 100-amp feed (from a breaker box) to five 20 amp, 120 V outlets with circuit breakers. The unit works on 120, 240, or 208 V three-phase circuits and provides spike and surge suppression. Furman's Website is www.furmansound.com.

Power Source

If your recording system is a multitrack recorder that will connect to the PA mixer's insert sends, simply set up next to the PA mixer and plug into the same AC outlet strip that the PA mixer is using.

If you split the mic signals, you can run your recording snake up to the front-of-house (FOH) position and set up your mixer or mic preamps there. Get AC power from the PA mixer's power outlet strip so that your mixer and the PA mixer have the same ground voltage. This prevents hum when the two mixers are connected. Of course, you might prefer to locate in an isolated area for better monitoring. If you plug into a local AC outlet there, you should be able to make connections without ground loops and hum if the house power distribution system is well designed. Otherwise, run one or two thick (14 or 16 gauge) extension cords from the PA mixer's outlet strip to your recording system. These cords may need to be 100–200 feet long. Plug AC outlet strips into the extension cord, then plug all your equipment into the outlet strips.

If you're using a remote truck, find a source of power that can handle the truck's power requirements, usually at a breaker panel. Some newer clubs have separate breaker boxes for sound, lights, and a remote truck. Find out whether you'll need a union electrician to make those connections (US only). Label your breakers.

Check that your AC power source is not shared with lighting dimmers or heavy machinery; these devices can cause noises or buzzes in the audio. You might want to use a power conditioner with an AC isolation transformer.

The industry-standard power connector for high-current applications in the US is the Cam-lok, a large cylindrical connector. Male and female Cam-loks join together and lock when you twist the connector ring. Distro systems and power cables with Cam-lok connectors can be rented from rental houses for film, lighting, electrical equipment, or entertainment equipment. One such rental house is Mole-Richardson (www.mole.com.). The standard power connector in Europe is the C-form.

Use an adapter from Cam-lok to bare wires. Pull the panel off the breaker box, insert the bare wires, and connect the Cam-lok to your truck's power. Some breaker boxes have Cam-loks already built in. (*Caution*: Have an electrician do the wiring if you don't know what you're doing. A union electrician might be required anyway.)

Measure the AC line voltage. If the AC voltage varies widely, use a line voltage regulator (power conditioner) for your recording equipment. If the AC power is noisy, you might need a power isolation transformer.

Check AC power on stage with a circuit checker. Are grounded outlets actually grounded? Is there low resistance to ground? Are the outlets of the correct polarity? There should be a substantial voltage between hot and ground, and no voltage between neutral and ground.

Some recording companies have a gasoline-powered generator ready to switch to if the house power fails. If there are a lot of lighting and dimmer racks at the gig, you might want to put the truck on a generator to keep it isolated from the lighting power.

Interconnecting Multiple Sound Systems

If you hear a hum or buzz when the systems are connected, first make sure that the signal source is clean. You might be hearing hum from a broken snake shield or an unused bass-guitar input. If the hum persists, experiment with flipping the ground-lift switches on the splitter and on the direct boxes. On some jobs, you need to lift almost every ground; on others, you need to tie all the grounds. The correct ground-lift setting can change from day to day because of a change in the lighting. Expect to make some trial-and-error adjustments.

If the PA has serious hum and buzz problems, offer help. You might hear buzzes in your quiet control room or isolating earphones that the PA people can't hear over the main system with noise in the background. Maybe the PA company is using balanced line-level audio cables that are grounded at both ends, which can cause ground loops and hum. You might want to use some cable ground-lift adapters (Figure 4-2) to float (remove) the extra pin-1 ground connection at equipment inputs.

A radio station or video crew might want to take an audio feed from your mixing console. If so, you can prevent a hum problem by using a console with transformer-isolated inputs and outputs. Or use an XLR Y-cable with a ground lift. Other options are a direct box or a line-level transformer splitter, such as the Whirlwind Line Balancer/Splitter (LBS) or the ProCo IT-1 Isolation Transformer Unit. For best isolation, use a distribution amp with several transformer-isolated feeds. Lift the cable shield at the input of the system you're feeding.

We recommend the article, "Sound System Interconnection" on the Rane Website, http://www.rane.com/note110.html. It describes how to connect balanced and unbalanced equipment and prevent ground loops.

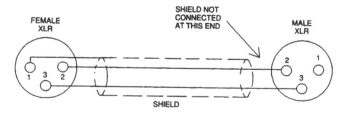

Figure 4-2 A ground-lift adapter for balanced line-level cables.

Mic Connections

You have previously created a mic list, so you know what to plug in where. Make some copies of the mic list. After unpacking, place one list by the splitter, another by each stage box, and another by each mixer. The list will act as a guide to help you keep things organized.

Attach a strip of white console tape just below the mixer faders. Use this strip to write down the name of the instrument that each fader affects.

Based on the mic list you made, you might plug the bass direct box into snake input 1, plug the kick mic into snake input 2, and so on. Label fader 1 "bass," label fader 2 "kick," etc. Also plug in equipment cables according to your block diagram.

Have an extra microphone and cable offstage ready to use if a mic fails.

If you unplug a mic plugged into phantom power, it will make a popping noise in the sound-reinforcement system. Be sure to mute the mic channel first.

Running Cables

To reduce hum pickup and ground-loop problems associated with cable connectors, try to use a single mic cable between each mic and its stage-box connector.

Avoid bundling together mic cables, line-level cables, and power cables. If you must cross mic cables and power cables, do so at right angles and space them vertically.

Plug each mic cable into the stage box or splitter (if used), then run the cable out to each mic and plug it in. This leaves less of a mess at the stage box. Leave the excess cable at each mic stand so you can move the mics. Don't tape down the mic cables until the musicians are settled.

It is important that audience members do not trip over your cables. In high-traffic areas, cover cables with rubber floor mats or cable crossovers (metal ramps). At least tape them down lengthwise with gaffers tape.

It helps to set up a closed-circuit TV camera and TV monitor to see what's happening on stage. You need to know when mics get moved accidentally, or when singers use the wrong mic, etc.

Setting Up the Recording Mixer

If you are using mic splitters and a recording mixer, here is a suggested procedure:

1. If the mixer is set up in a dressing room or locker room, add some acoustic absorption to deaden the room reflections. You might bring a carpet for the floor plus acoustic foam or packing blankets for the walls.
2. Turn up the recording monitor system and verify that it is clean.
3. Plug in one mic at a time and monitor it to check for hums and buzzes. Troubleshooting is easier if you listen to each mic as you connect it, rather than plugging them all in and trying to find a hum or buzz.
4. Check and clean up one system at a time: first the sound-reinforcement system, then the stage-monitor system, then the recording system. Again, this makes troubleshooting easier because you have only one system to troubleshoot.
5. Use as many designation strips as you need for complex consoles. Label the input faders at the bottom and top. Also label the monitor-mix knobs and the meters.
6. Verify that left and right channels are correct and that the pan-pot action is not reversed audibly.
7. If you are setting up a separate recording monitor mix, do a preliminary pan-pot setup. Panning similar instruments to different locations helps you identify them.
8. Make a short test recording and listen to the playback.

Mic Techniques

Normally the PA company chooses and places the mics, but you might do it yourself or collaborate with the PA company. This section offers some tips on miking instruments and vocals.

In a quiet recording studio with good isolation between instruments, you have the freedom to mike instruments 1 or 2 feet away if you want. But in a noisy club or auditorium, with band members close together on stage, separation is a serious problem. Usually you need to mike a few inches away to reduce background noise, room acoustics, leakage, and feedback. Here are some other ways to control these problems:

- Use directional microphones, such as cardioids, supercardioids, or hypercardioids. These mics pick up less feedback, leakage, and noise than omnidirectional mics at the same miking distance. For example, the Audix OM5, OM6, and OM7 vocal mics (www.audixusa.com) have a tight hypercardioid pattern that controls leakage.

- If possible, hang drapes or other acoustical absorption material on the rear wall to reduce sound reflections into the mics. Consider moving the instruments and vocals farther apart to improve isolation and reduce phase interference between mics.

- Use direct boxes. A direct box (DI box) is an interface between an unbalanced electric instrument and a balanced mixer mic input. Bass guitar, electric guitar, and keyboards can be recorded direct to eliminate leakage and noise in their signals. However, nothing beats the sound of a miked guitar amp. You could record the guitar directly from its effects boxes, then use a guitar-amp emulator during mixdown. That is especially helpful if the guitar amp is noisy. Note that sequencers and some keyboards have high-level outputs, so their direct boxes need transformers that can handle line level.

- Use contact pickups. On acoustic guitar, acoustic bass, and violin, you can avoid leakage by using a contact pickup. It is sensitive only to the instrument's vibration, not so much to sound waves. The sound of a pickup is not as natural as a microphone, but a pickup may be your only choice. Consider using both a pickup and a microphone on the instrument. Connect just the pickup to the PA and monitor speakers to prevent feedback, and connect the mic and pickup to the recording mixer. If the mic has too much leakage during mixdown, you can either use the pickup track or overdub the guitar.

- Choose mic positions carefully. Close miking affects the tonal balance of a recorded instrument. When you change the mic position, you change the tone quality. For example, an acoustic guitar miked near the sound hole is bassy, near the bridge is mellow, and near the fingerboard is bright.

Listed below are some typical miking methods for vocals and instruments. These are just some suggested starting positions—experiment and use your ears.

Vocal: Use a condenser, ribbon, or dynamic vocal mic about 0–3 inches from the mouth. To reduce breath pops with vocal mics, be sure to use foam pop filters. Leave a little spacing between the pop filter and the mic grille. It also helps to switch in a low-cut filter (100 Hz high-pass filter). If the mic is a cardioid, aim the "dead" rear of the mic at the floor monitors to reduce feedback. If the mic is a supercardioid or hypercardioid, angle the mic to be more nearly horizontal so that its zone of least pickup aims at the monitors. Caution the performer not to cover the grille with a hand because it could color the sound and cause feedback.

Acoustic guitar: Option 1—Mount a mini mic at the edge of the sound hole and turn down the excess bass with the mixer's equalization (EQ). Option 2—Mount a mini mic inside the guitar under the 17th fret. Option 3—Mount a mini cardioid mic a few inches from where the fingerboard joins the body, or place a boom-mounted mic there. Option 4—If you can't get enough gain-before-feedback with a miked acoustic guitar (as often happens with a rock band), use a pickup or a pickup mixed with a mic. Plug the pickup into a direct box. During mixdown, you might roll off some highs around 10–12 kHz to make the pickup track sound less "electric."

Sax: Place a dynamic or condenser mic a few inches above the bell, aiming at the tone holes. For more mobility and a brighter sound, clip a mini mic to the bell.

Electric guitar or electric bass direct: For a clean sound, plug the guitar into a direct box. Plug the direct-box output into a mixer mic input. For a distorted sound, plug the instrument into a guitar signal processor, and connect the processor output into a mixer line input.

Electric guitar amp: Place a dynamic mic with a presence peak 1 inch from a speaker cone, slightly off-center.

Synthesizer or drum machine: Use direct boxes. Flip the ground-lift switch on the boxes to the position where you monitor the least hum.

Drum set: Try two cardioid condenser "stick-type" mics about 2 feet above the cymbals (see Chapter 8 for stereo miking techniques). Add a dynamic mic or mini condenser mic just above the snare-drum rim, and a large-diaphragm dynamic mic in the kick drum. Put some dynamic mics just inside the tom-tom rims if necessary. Stuff a pillow or blanket in the kick to get a tight sound, and use a wood or plastic beater for extra "click." The kick drum often requires some EQ to sound good. You might cut a few dB around 400 Hz and boost around 4 kHz.

Bongoes or congas: Try a dynamic mic midway between the drums a few inches away, or mike both drums up close.

Grand piano: Raise the lid. Gaffer-tape a mini mic or boundary mic to the underside of the lid in the middle. For stereo, use two mics: one over the bass strings and one over the treble strings. If you need more isolation, close the lid and tweak EQ to remove the tubby coloration (usually cut 1–4 dB around 250 Hz).

Another method: raise the lid and place two condenser mics 8 inches over the bass and treble strings, about 8 inches horizontally from the hammers. Or place the bass mic about 2 feet toward the tail. If necessary, cut 1 or 2 dB around 250 Hz to reduce tubbiness.

Upright piano: Face the soundboard toward the room (not next to a wall). Mike the soundboard a few inches from the bass and treble areas with two dynamic or condenser mics.

Banjo: Tape a mini omni mic to the drum head about 2 inches in from the rim, or on the bridge. Or place a cardioid dynamic or condenser mic about 6 inches from the drum head.

Fiddle: Try a dynamic, ribbon, or condenser mic about 8 inches over the top. For a fiddle player who sings, place the mic about 6 inches over the fiddle, aiming at the player's chin. Some players use a pickup into a direct box.

Mandolin, bouzouki, dobro, lap dulcimer: Place a dynamic, ribbon, or condenser mic about 6 inches away.

Hammered dulcimer: Place a dynamic, ribbon, or condenser instrument mic about 8 inches above the top edge aiming at the soundboard.

Acoustic bass: Mike about 3–6 inches away, just above the top of an f-hole or a few inches above the bridge. For more isolation, wrap a cardioid dynamic mic in foam (except for the grille) and stuff it behind the bridge. Many players use a pickup, which they plug into a bass-guitar amp. Plug the pickup into a direct box and connect the phone-jack output in the direct box to the amp. Some amp heads have an XLR direct out and a ground-lift switch to prevent hum. The pickup will need some EQ to sound natural.

Flute: Place a dynamic, ribbon, or condenser mic halfway between the mouthpiece and the tone holes, about 6 inches away. Or wear a headworn mic, and place the mic capsule between the mouthpiece and tone holes.

Harmonica: Mike up close with a dynamic vocal mic. Or plug a hand-held mic into a guitar amp, and mike the amp.

Accordion, concertina: Tape a mini mic onto the tone holes on each side (two mics total). You might prefer to locate a dynamic mic 6 inches from the tone holes on the keyboard side. Some players use a pickup into a direct box.

When you're recording a band that has been on tour, should you use its PA mics or your own mics? In general, go with its mics. The artists and PA company have been using their mics for a while and may not want to change anything. Most mics currently used in PA are good quality anyway, unless they are dirty or defective.

If you're not happy with their choice, you could add your own instrument mics. Let the PA people listen to the sound in the recording truck, or in headphones. If it sounds bad because of the mic choice, ask, "Would it be okay if we tried a different mic (or mic placement)?" Usually it's all right with them—it's a team effort. Make sure every instrument is miked. If not, add your own microphones.

Electric-Guitar Grounding

While setting up mics, you need to be aware of a safety issue with the electric guitar. Electric-guitar players can receive a shock when they touch their guitar and a mic simultaneously. This occurs when the guitar amp is plugged into an electrical outlet on stage, and the mixing console (to which the mics are grounded) is plugged into a separate outlet across the room. If you're not using a power distro, these two power points may be at widely different ground voltages. So a current can flow between the grounded mic housing and the grounded guitar strings.

Caution: Electric-guitar shock is especially dangerous when the guitar amp and the console are on different phases of the AC mains.

It helps to power all instrument amps and audio gear from the same AC distribution outlets. If you lack a power distro, run a heavy extension cord from a stage outlet back to the mixing console (or vice versa). Plug all the power-cord ground pins into grounded outlets. That way, you prevent shocks and hum at the same time.

If you're picking up the electric-guitar direct, use a transformer-isolated direct box and set the ground-lift switch to the minimum-hum position.

Using a neon tester or voltmeter, measure the voltage between the electric-guitar strings and the metal grille of the microphones. If there is a voltage, flip the polarity switch on the amp (if any). Use foam windscreens for additional protection against shocks.

Audience Microphones

When you make a live recording, audience-reaction mics are essential. They help the recording to sound "live." Without audience mics, the recording may sound too dry, as if it were done in a studio. And there's nothing like cheering and clapping to add excitement to a live recording.

One easy method is to aim two good cardioid condenser mics at the audience. Hypercardioid or shotgun mics are even better. Put them on regular mic stands, on the stage floor, on either side of the stage. If those

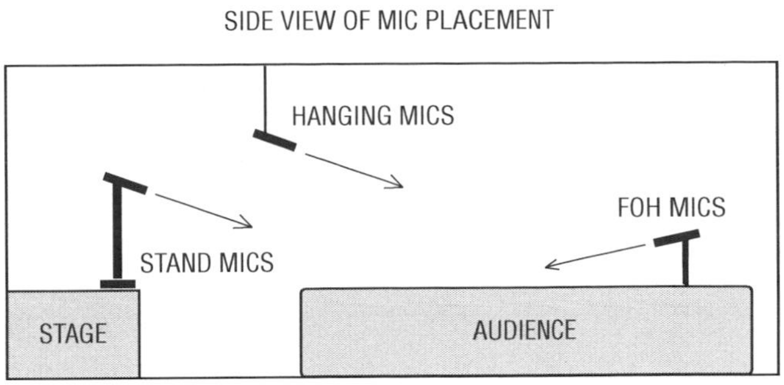

SIDE VIEW OF MIC PLACEMENT

Figure 4-3 Some audience miking techniques.

mic stands must not be seen, hang some mics or put a stereo pair at FOH (Figure 4-3). Chapter 8 describes stereo mic techniques.

If the audience mics are far back in the hall—100 feet from the stage, or at FOH, for example—they pick up the band's sound with a delay. When mixed with the close mics, the audience mics add an echo. You can prevent this echo if you mix the recording using computer software: slide the audience tracks to the left (earlier in time). Align the waveforms of big peaks.

What if you don't have enough tracks for the audience mics? Record them on a two-track recorder. Load this recording into your digital audio workstation (DAW) along with the other tracks. Align the two recordings in time as just described. Another option is to turn up some stage mics in the mixdown when the audience applauds.

If the audience mics are run through the PA mixer's preamps, leave the audience mics unassigned in that mixer to prevent feedback.

To get more isolation from the house speakers in the audience mics, use several mics hung close to the audience. Some engineers put up 4 audience mics maximum; some use 8 to 10. Use directional mics and aim them away from the house speakers.

Another option is to *not* mike the audience or not use the audience tracks. Instead, during mixdown, you could simulate an audience with audience-reaction CDs. Simulate room reverb with an effects device or plug-in.

Setting Levels and Submixes

Now that the mics are set up, you might have time for a sound check. That's when you set recording levels. Have the band play a loud song.

Locate a mixer input module that is feeding a recorder track. Set the input trim (mic preamp gain) to get the desired recording level on each track. You might set each track's level to peak around −10 dB maximum, which allows some headroom for surprises.

Check all the keyboard patches and guitar effects because some may be much louder than others. You might put a limiter in line with some insert sends to prevent excessive levels.

You often encounter PA consoles where some insert sends are tied up with signal processors. You must use those channels' direct-out jacks instead, which usually are postfader (unless they can be switched to pre-fader). Another option is to "Y" the insert send to your recorder and to the processor input. Or, assign those channels to unused groups (buses) and get your recording signals from there.

If you don't have enough tracks for all the mics, you could set up a two-track drum submix. Ideally you would do this with the recording mixer rather than the PA mixer, and monitor over headphones or Nearfield monitors in a separate room. Assign each drum mic to buses 3 and 4 (for example), and pan each mic as desired. Put the submaster faders for buses 3 and 4 at design center—the shaded area about 1/2 to 3/4 up.

Have the drummer hit each drum repeatedly, one at a time, as you adjust the input trims to prevent clipping. For example, ask the drummer to bang on the kick drum. Turn down the kick-drum's input trim all the way. Slowly bring it up until the clip LED (overload light) flashes. Then turn down the input trim about 10 dB to allow some headroom. When all the drum trims are set, do a drum mix with the faders. Try to adjust all the faders up or down by the same amount so that the recording level is correct when the submasters are at design center.

Recording

If your recording will be synched later with a video recording by using SMPTE (Society of Motion Picture and Television Engineers) time code, record the video time-code feed on a spare recorder track.

A few minutes before the band starts playing, start recording. Keep a close eye on recording levels. If a track is going into the red, slowly turn down its input trim and note the counter time where this change occurred.

Caution: If you are recording off the PA mixer, turning down its input trim will affect the PA levels. The PA operator will need to turn up the corresponding monitor send and channel fader.

This is a touchy situation that demands cooperation. Ideally, you set enough headroom during the sound check so you won't have to change levels. But be sure the PA operator knows in advance that you might need to make changes. Ask the operator whether he or she wants to adjust the gain trims for you, so the operator can adjust corresponding levels at the same time. Thank the operator for helping you get a good recording.

If you are recording with a splitter and mic preamps on stage, assign someone to watch the levels and adjust them during the concert. Preamps with meters allow more precise level setting than preamps with clip LEDs.

Keep a track sheet and log as you record. For each song in the set list, note the counter time when the song starts. Later, during mixdown, you can go to those counter times to find songs you want to mix. Also note where any level changes occurred so you can compensate during mixdown. It helps to note a counter time when the signal level was very high. When you mix the recording you can start at that point in setting your overall mix level.

Record each set nonstop so that you don't miss anything. You can edit out unwanted material later.

Teardown

After the gig, because your mics might be stolen or damaged, pack them away first. Refer to your equipment list as you repack everything. Note any equipment failures and fix broken equipment as soon as possible.

After you haul your gear back to the studio, it's time for mixing and editing, covered next.

5

AFTER THE SESSION: MIXING AND EDITING

The gig or concert is over, and you've brought your recording back to the studio. First, be sure to back it up to a separate hard drive. If you recorded live to two tracks, you are ready to dump the recording to your computer and edit it for CD release. If you recorded to multitrack, you will mix the tracks to stereo, then edit the stereo mix. This chapter describes the process.

Editing a Two-Track Recording

Suppose you recorded a gig in stereo directly to two-track. Below is a procedure to get the recording into your computer and edit it. This will let you prepare a CD of your recording:

- If you used a flash-memory recorder or MP3 player/recorder, connect its USB (Universal Serial Bus) port to the one on your computer. The computer will detect the recorder as a data-storage device. Click-drag the wave file(s) from the recorder to the hard drive that you use for audio files. If you recorded several wave files—one for each musical set—load them all in.

- If you used a Hi-MD MiniDisc recorder, use Sony's SonicStage software to copy the wave file(s) to the hard drive via USB.

71

- If you used a digital audiotape (DAT) recorder, plug its line outputs into your sound card's line input. Play the gig recording, set the recording level, and copy it in real time to your computer. If your sound card has a digital input, connect that to the DAT recorder's digital output for higher quality. Make sure the sampling rate of your software matches that of the DAT recording. Set the DAT as the master clock.
- If you used a four-tracker, set up a pleasing mix of all the tracks. Plug the stereo line outputs of the four-tracker into the sound card's line input. Play the gig recording, set the recording level, and copy the stereo mix in real time to your computer.

Now you can launch your recording software and open the gig's wave file that is on your hard drive. Here is a simple method to create tracks for a CD, one track per song, with silent spaces between the songs:

1. Start with one long wave file of the entire set. It is a single clip, region, or segment of audio. Each song appears as a group of high-level waves, and the quiet pauses between songs are low-level waves (Figure 5-1). Normalize the file so that it will be as loud as possible without distortion.

2. Split that long, single clip into separate clips, one per song. Keep a few seconds of crowd noise and applause before and after each song.

3. Cut out unwanted sounds and pauses between songs. Then your computer screen will look something like Figure 5-2.

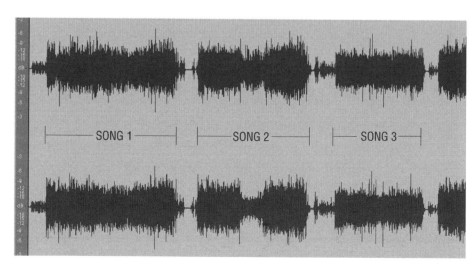

Figure 5-1 Songs appear as high-level waves in the set's sound track.

4. If a song is too long to sustain interest on a CD, remove verses, choruses, or solos to shorten it.

5. Zoom way in to the beginning of the song. Add a fade-in before the song begins (Figure 5-3). During the applause after the song, add a fade-out with a cosine shape (as in Figure 5-6).

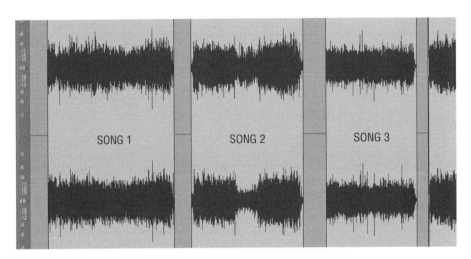

Figure 5-2 Edited song clips.

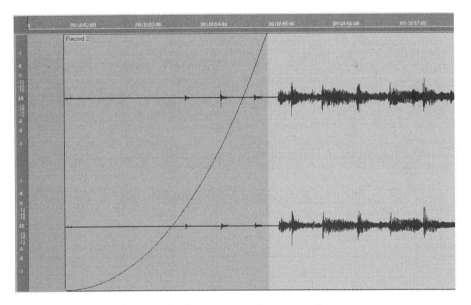

Figure 5-3 A fade-in at the beginning of a song clip.

6. Next you will save each song's clip as a separate wave file. Depending on your software, either (1) highlight a clip and export it to a file or (2) place song clips on separate tracks, solo one track at a time, and export it to a file. Name each file the same as the song title.

7. Finally, launch your CD-burning software, load the wave files into a playlist, and burn a CD-R.

Other ways to create albums and demos from stereo recordings are described later under the headings "Mastering an Album" and "Mastering a Demo."

Preparing to Mix a Multitrack Recording on a Computer

Suppose you brought a multitrack recording back to the studio, and you want to mix it to stereo. Mixing can be done with a hardware mixer or with computer digital audio workstation (DAW) software. This section focuses on the DAW process.

If you made the multitrack recording directly to a computer, you're ready to mix. Skip to the next section.

If you recorded on a multitrack hard drive recorder, and you want to mix the tracks with your computer, first copy the tracks to your computer's hard drive. Depending on the recorder model, use TDIF, Lightpipe, or Ethernet cables for the audio transfer to computer. If you used an Alesis HD24 recorder, you can put its hard drive into an Alesis Fire-Port interface and copy the files to your computer.

Split the Gig Recording into Song Projects

At this point, you need a way to split the multitrack gig recording into separate multitrack song recordings. In other words, create one multitrack recording project for each song so that you can mix each song separately. Here are the steps:

1. Launch your DAW recording/editing software. Load a multitrack template. Import all the tracks from Set 1. You should see the track waveforms on your computer monitor (Figure 5-4). The tracks will be as long as the set was, typically 40 minutes. Select **File** > **Save as**, and name the project "Set 1."

2. Select **File** > **Save as**, and this time name the project "Song 1" (where "Song 1" is the title of the first song you want to work on).

3. Play the tracks and find the beginning of Song 1. Put the cursor a few seconds before Song 1 where you hear some crowd noise.

4. Select all tracks, then split all of them at that point (Figure 5-4, the left side of the dark area). Use the "split tracks" (or similar) command in your software to do that.

5. Select all the tracks to the left of the split point and delete them, closing up the space that is left.

6. Put the cursor a few seconds after the applause following Song 1.

7. Select all tracks, then split all of them at that point (Figure 5-4, the right side of the dark area).

8. Select all the tracks to the right of the split point and delete them. Only the tracks for Song 1 remain.

9. Again, save the project as "Song 1" or whatever. You have created a separate multitrack project for Song 1.

10. Now you will repeat this process for the rest of the songs. Load the Set 1 project. The tracks for the entire set appear. Repeat Steps 2–9 for Song 2, Song 3, and so on.

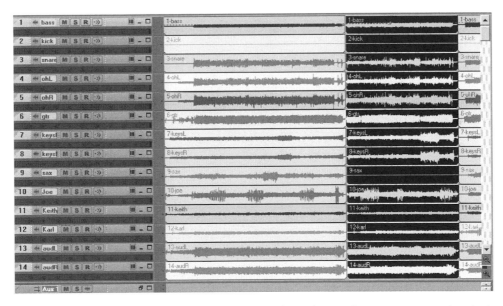

Figure 5-4 Several tracks loaded into a multitrack template on screen. The dark area is the group of tracks for one song.

Now, when you load the project called Song 1, the multiple tracks for Song 1 will appear on screen. They will be as long as Song 1 was. You can mix and edit those tracks, and export the mix for Song 1. The same goes for Song 2, Song 3, and so on.

Delete Unwanted Material

Now you can edit Song 1. Listen to each track by itself or look at its waveform on your computer monitor. Erase or delete anything you don't want to keep. For example, erase some unwanted talking on the vocal track during the song intro. Or erase a thump on a guitar track that happened when someone bumped into the mic stand.

Preparing to Mix a Multitrack Recording with a Mixer

If you recorded on a multitrack hard drive recorder and you want to mix the tracks with a hardware mixer, you're ready to go. First, remember that live gig recordings are long, nonstop programs. It's hard to mix an entire set continuously for an hour or so without making a mistake. A better procedure is to set up a mix for the first song, record it or export it to your hard drive, set up a mix for the next song, record it, and so on. Then edit the mixes together to sound like a single concert.

When mixing each song, be sure to include several seconds of crowd noise or applause before and after each song. You will use this crowd sound when editing later. **Don't fade in or fade out while mixing; you will do the fades while mastering**.

Do Punch-Ins

Some band members might want to punch-in (re-record) parts of tracks to correct mistakes or to replace tracks that have too much leakage (usually vocals or a miked acoustic guitar). Be careful to match the sound of the instrument in the studio to the sound of the track. Use the same mic and mic position that you used during the gig. You might want to add some reverb to the re-recorded section because otherwise it will sound dry, unlike the live-recorded tracks.

Mix Each Song

Mixdown procedures can be found in recording textbooks, so I won't cover them here. On-location recordings, however, have some unique mixing challenges.

Fader levels: Start with the loudest part of the recording. You noted the counter time of this event when you wrote the log during the gig. Set fader levels so that the stereo mix peaks at about −3 dBFS maximum when the master faders are at 0 dB. The more tracks in use, the lower you need to set the fader levels to prevent clipping the stereo output bus.

Set up fader automation so that each solo comes in at an appropriate level. Another use for automation is to bring up the applause when needed.

EQ (equalization): If background noise and leakage are problems, use EQ to filter out frequencies above and below the spectrum of each instrument. For example, cut below 100 Hz on vocal tracks. This helps to reduce breath pops and bass leakage into the vocal mics. Also use EQ to compensate for the unnatural tonal balances you can get with close miking or direct boxes. Roll off excess bass on a close-miked vocal or acoustic guitar, and so on.

Panning: You might want to pan the tracks to match the locations of the instruments and vocals at the gig. Try to achieve a good balance between left, center, and right instruments. For example, you might pan stereo keyboards toward the left and lead guitar to the right. Bass, kick, snare, and lead vocal are usually panned to center.

Compression: Use compression on some individual tracks to control the extreme level variations that may occur in live recordings (especially on vocals).

Gating: Use a noise gate to remove buzzes between electric-guitar passages, or to reduce leakage in the tom, snare, and kick-drum tracks.

Reverb: If the tracks sound too dry, as if they were done in a studio, add a little reverb with a decay time like that of the venue. Or mix in the audience tracks if they sound good. Apply the same stereo reverb device (or plug-in) to all the tracks so they sound like they are in the same room. To do that, set up an aux send on each track, with all the sends feeding the same reverb bus. Omit the reverb on the kick and bass to prevent muddiness. Turn down the reverb during spoken introductions.

Echo: Any echo applied to a vocal track will appear on any instrument that leaked into the vocal mic. For example, you might hear an echo on the drum leakage. Avoid echo if you have this problem.

Pitch correction: If the mics picked up a lot of leakage from the PA or monitor speakers, any pitch correction applied to the vocal track might sound strange. The vocal track will be pitch-shifted, but the leakage won't, so you will hear two pitches.

Audience tracks: Keep the audience mics in the mix to make it sound live. Audience mics can muddy the sound if mixed in too loudly. Keep them down in level, just enough to add some ambience. Bring them up gently to emphasize crowd reactions. You can reduce muddiness by rolling off the lows in the audience mics.

The audience mics might sound delayed compared to the stage mics, causing an echo. If so, slide the audience tracks earlier in time so that their waveforms align with those of the stage mics.

Create a template: Once you've set up a good mix for Song 1, save your settings as a template that you can recall for all of the other songs.

CD track 26 demonstrates a mixdown of a blues band live recording. You will hear the effects of panning, EQ, compression, reverb, gating, and audience mics.

Mixing for Surround Sound

If you are mixing a live gig for surround, you can put the listener inside a simulated concert hall. To do this, feed the audience mics to the rear surround channels. Also feed some or all of the reverberation to the rear, whether it was miked or made with an effects unit.

To create the subwoofer channel, turn up an aux send equally on each input channel that has low-frequency content. Low-pass filter the aux-mix output at 120 Hz. That filtered aux becomes your low-frequency effects (LFE) channel. You also could use a spare bus for the sub-channel and assign bass, kick, and low-synth notes to that bus.

To adjust the overall monitor level of the six tracks, you could plug them into a surround receiver and use its volume control. Or, if you have an automated console, you could control six input modules (used only for monitoring) with a single group fader.

A monitor level that does not exaggerate the bass and treble is 85 dB SPL. Get an SPL meter (say, from Radio Shack) and set it to C-weighted, slow-reading scale. Play pink noise through your monitor speakers one at a time, and set each speaker's monitor level to get an 85 dB SPL reading.

In stereo mixdowns it's common to pan the bass, kick drum, snare, and lead vocal to center. But feeding all these sounds to the center speaker can cause it to overload. You might want to feed these sounds partly to front-left and front-right, as this takes some of the load off the center speaker.

If you start with an instrument in the center speaker, you can send some of its signal to the surrounds to move the instrument toward you.

A center *phantom image* is an apparent source of sound in the center of two stereo loudspeakers. The sound appears in the center when the level and timing of the signals in both speakers are equal. A center speaker has more output around 2 kHz than a center phantom image does, because the phantom image has some phase cancellation at 2 kHz due to sound delay around the head. The 2 kHz bump in the center speaker can sound annoying on vocals, so be careful with EQ.

Instruments are more distinct in surround mixes. It's easier to hear what each one is doing. So you may find that you need less level-adjusting and less compression in surround than in stereo.

You might start with a standard stereo mix on two speakers, then add the center channel, surrounds, and sub. Adjust the mix to send some tracks to the other speakers.

Be sure to check your surround mix in stereo and mono to make sure it is compatible. The perceived reverb level in stereo listening should match the perceived reverb level in multichannel listening. Most engineers prefer to do separate stereo and surround mixes, rather than doing a surround mix and hoping it sounds good when folded down to stereo.

A suggested channel assignment for surround mixes is:

1. Left front
2. Right front
3. Center
4. LFE (or subwoofer channel)
5. Left surround
6. Right surround
7. SMPTE (Society of Motion Picture and Television Engineers) time code (if desired).

Mastering an Album

At this point, your mixes are recorded on your hard drive. Each song mix is a separate wave file. Now you can master the mixes with recording/ editing software. When the mastering is done, the songs will play in the desired order with a smooth transition between them. There will be no excess pauses or noises. Let's run through the process step by step:

1. Launch your recording software and open a multitrack template. Import the first song's mix into track 1. Import the second song's mix into track 2 at the end of the first song. Repeat for the rest of the songs on successive tracks (Figure 5-5). Because each song is on its own track, it's easy to change the level or EQ for each track and slide song clips in time.

2. The audio waveform of each song appears on your monitor screen. You can zoom out to see the entire program, or zoom in to work on tiny spans of time.

3. At the beginning and end of each mix, edit out any unwanted sounds or pauses. There should be some crowd noise or applause before the first note. That way, you can start the edited recording by fading up on crowd noise. This helps to establish the recording as being "live."

   If there is no crowd noise before the first song, find some crowd noise elsewhere in your recording. Mark off a section of it about 5 seconds long and save it as a region or clip called "Crowd."

4. Is any song too long to sustain interest on a CD? Maybe that 15-minute jam worked at the live show, but it might get boring with repeated home listening. Remove verses, choruses, or solos to shorten the song.

5. Slide Song 2 in time so that the end of Song 1's applause overlaps the beginning of Song 2 (Figure 5-6). Or overlap the crowd noise after Song 1's applause with the beginning of Song 2. You might want to fade out the end of Song 1's clip (Figure 5-6). Repeat for the rest of the songs. If done well, the edited program should sound like a single continuous concert.

6. You might need to adjust the level of each track so the songs are equally loud. Also, you might want to add EQ or Harmonic Balancing to songs that need it. If you want to make a hot (loud) CD, set the output of each track to a common stereo bus. Insert a peak limiter/normalizer in that stereo bus, and adjust its threshold for the desired amount of limiting.

7. Once you're happy with the program, write down the start time of each song. You will use those start times to create a cuelist that creates track start ID's when you burn a CD. Some CD-burning software uses marks on a waveform to define the start ID's.

Figure 5-5 Placing each song clip on its own track makes mastering easier.

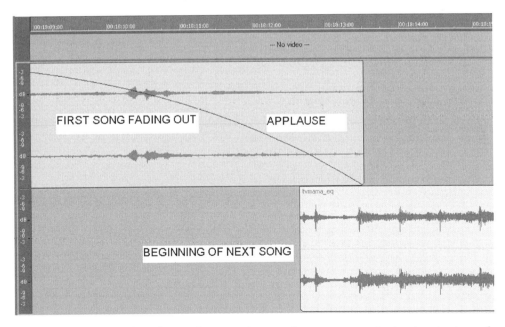

Figure 5-6 Overlap the ending applause of one song with the beginning of the next song.

8. Finally, turn on dither and export the mix to a 16-bit stereo wave file. That single file contains all the songs in the album with no silence between them.

Time to burn a master CD. You need CD-burning software that lets you burn a CD from a cuelist or cuesheet. This is a list of start times for all the songs within the single stereo wave file you just exported. It tells the CD burner when to start each CD track. The cuelist also gives the filepath of the mastering wave file. One example of cuelist-controlled software is CdRWIN, available from www.goldenhawk.com.

In the cuelist, make each song's start time 10 frames earlier than the actual start time. That way, during CD playback no audio will be skipped at the beginning of each track. After you type the cuelist, burn a CD. There's your finished CD master.

Mastering a Demo

If you want to create a demo CD rather than an album, you might want to keep only about 1 minute of each song. That way, the listener can quickly sample all the band's styles without having to hear each song from start to finish. The edited version would be something like this:

Song 1 intro, verse, chorus, fade out.

Song 2 intro, verse, chorus, fade out.

Song 3 intro, verse, chorus, fade out.

And so forth.

Last song intro, verse, LAST chorus, and fade out applause.

You could fade into the middle of each song if that is more effective than starting each song from the beginning.

Here is a procedure to edit the song mixes to create song samples. Let's say you imported all the song mixes to the tracks as in Figure 5-5. Each song mix appears as a clip on a separate track:

1. Place the cursor a few seconds before the first song and edit out the audio before that point. In other words, trim the beginning of the first song.

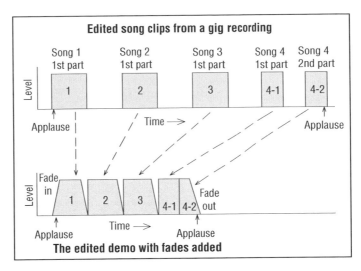

Figure 5-7 (Top): Edited song clips from a gig recording. (Bottom): The edited demo with fades added.

2. Place the cursor about 5 seconds after the first chorus ends, and split the clip there. Delete the rest of the song to the right of the split point.

3. For each remaining song, delete the pause just before the beginning of the song (leaving a space), and delete the rest of the song starting about 5 seconds after the first chorus ends. You will end up with a group of approximately 1-minute clips, one clip per song (Figure 5-7, top).

4. You want the demo to end with applause. So, when you edit the last song, you will keep the ending of the last song and about 15 seconds of applause afterward. By "last song," I mean the last song you want to play in your edited program. It's not necessarily the last song played at the gig.

Display the waveform of the last song clip. Using the DAW editing tools, split the clip at the beginning of the first chorus and at the beginning of the last chorus (Figure 5-8). Zoom way in so you can mark those points with precision, just before a beat. Cut out the audio between the split points and close up the space.

Play the transition from part 1 to part 2 in the last song. If your timing was correct, it should sound like a single, nonstop song. If not, you can go back and fine-tune the edit points.

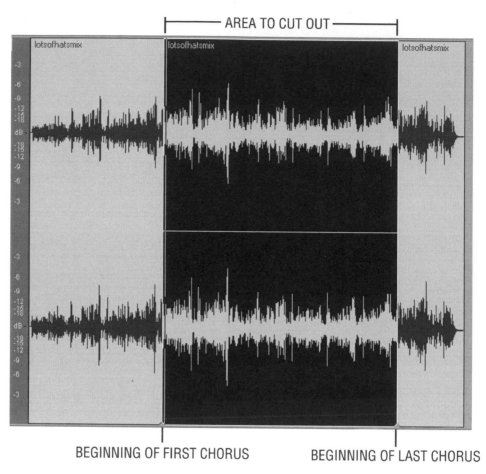

Figure 5-8 Cutting out part of the last song.

Add Fades and EQ

Now that you edited the songs into short samples, you can put them close together and add the fades. First, slide the song clips together so that Song 2 starts right after Song 1 ends, Song 3 starts after Song 2, and so on (Figure 5-7, bottom):

1. Using your DAW's fade function, fade up the applause or crowd noise just before Song 1. Fade up over 4 seconds or so (Figure 5-7, bottom left).

2. At the end of Song 1's chorus, fade to silence over about 4 seconds.

3. Immediately after Song 1 fades out, start Song 2 at full volume. Fade the end of Song 2, start Song 3 at full volume, and so on (Figure 5-7, bottom).

4. At the end of "Last song part 2," let the applause run for about 3 seconds, then slowly fade it out over about 8 seconds.

Those fade times are just suggestions—use your ears and do what sounds right.

When you're finished, export the mix to a stereo wave file. Burn a CD master from that file, and enjoy your demo.

6

A REAL-WORLD EXAMPLE: RECORDING A BLUES BAND IN A CLUB

In this chapter I'll describe an on-location recording project that I did recently. The leader of a local blues band phoned me, saying that his band wanted to be recorded live in a club. The band members thought that a live recording would capture their energy better than a studio recording, and would cost less to make.

Preproduction

In talking with the leader, I got this information:

Performance date and time: May 6, 2005, at 8 p.m.

Load-in time: 5:30 p.m.

Length of performance: Three 40-minute sets (approximately).

Venue and venue address: Riverside Café, Chester, Michigan. It is a restaurant with a stage for a band.

Instrumentation: Bass, drum set, electric guitar, two stereo keyboards (fed into a keyboard mixer), sax, and three vocals.

Directions to venue: I used Mapquest (www.mapquest.com).

In order to get a good kick sound with lots of attack, I asked the drummer ahead of time to remove the front head of the kick drum, use a wooden beater, and put a blanket in the bottom of the kick, pressed against the beater head.

I planned to do the PA mix and recording at the same time. I would mix the show over the band's PA speakers while recording each instrument's signal on a separate track for later mixdown back in the studio. The insert sends from the PA mixer would connect to a multitrack harddrive (HD) recorder.

Based on the instrumentation, I drew a block diagram of the recording system. From the block diagram, I generated a track list and equipment list:

Track	Instrument	Mic
1	Bass	Direct box
2	Kick drum	AKG D112
3	Snare	Shure SM57
4	Overhead L	Neumann KM140
5	Overhead R	Neumann KM140
6	Lead guitar	Shure SM57
7	Keyboards L	Direct box into keyboard mixer L output
8	Keyboards R	Direct box into keyboard mixer R output
9	Sax	Studio Projects B1
10	Lead vocal	Crown CM-200A
11	Vocal	Crown CM-200A
12	Vocal	Crown CM-200A
13	Audience L	Crown CM-700
14	Audience R	Crown CM-700

Other equipment

10 mic stands (three of these belonged to the band)
Stereo mic-stand adapter (for the audience mics)
14 mic cables
16-channel snake
16-channel Mackie 1604 VLZ mixer
Alesis HD24XR 24-track HD recorder
Snake from mixer insert jacks to HD recorder inputs

Headphones
Cable from the mixer's monitor send output to the band's monitor power amp
Cable from the mixer's master output to the band's PA power amp
AC output strip
Notebook, pens, gaffers tape, flashlight, and console tape

The Recording Session

On the day of the gig, I checked all the equipment and packed it in my car. I arrived at the venue 2½ hours before showtime. The venue owner showed me where I could set up (on a table near the stage). After the load-in, I set up the equipment according to the system block diagram. Gaffers tape kept the snake in place, so that no one would trip on it. The mixer insert sends fed the multitrack HD recorder inputs. I programmed the HD recorder to record 16 tracks, 24-bit, 44.1 kHz.

When the band arrived and set up, I placed the mics and plugged in the direct boxes. The audience mics were two cardioids mounted as a near-coincident stereo pair, on a mic stand near the mixer, aiming at the audience.

Next we did a sound check. I set recording levels using the mixer's gain trim pots. A minute before the band started playing, I hit Record. What a great show! I mixed it for the audience with the mixer's faders and equalization (EQ). This did not affect the levels or sound being recorded because the mixer's insert sends are pre-fader and pre-EQ.

While the band played, I watched the recording levels and noted the counter times where each song started and stopped. I also noted the time of the loudest part of the show. That would be a good place to set initial levels while mixing the multitrack recording later.

After the gig, I packed up, drove all the gear back home, and set it back up in the studio.

Preliminary Mix

The band wanted to hear a rough mix of the gig, so that they could decide which songs to include on the CD. I played the multitrack HD recording through an analog mixer, set up a mix, and recorded it nonstop to my computer's hard drive. Each 40-minute set was mixed to a separate wave file.

Looking at each file's waveform on-screen, I wrote down the time where each song started. I typed a cuesheet that listed those start times, then loaded the cuesheet into Cdrwin CD-burning software. Finally, I burned three CDs and mailed them to the band for evaluation.

Preparing for the Final Mixes

I planned to mix the recording in my computer digital audio workstation (DAW), so I needed to transfer the HD multitrack recordings to the computer hard drive. For each set, I copied the 14 tracks from the HD recorder into my computer by using an Alesis FirePort. That interface connects the recorder's hard drive to a computer's FireWire port for transferring audio tracks. On my D: drive, I created one folder per set to hold the tracks. Each set's folder contained 14 wave files, one per track, each about 40 minutes long.

Now I could split the long gig recording into shorter songs. I loaded all the tracks into a multitrack template in Cakewalk Sonar Producer and saved the project as Set 1. Then I split the wave files across all the tracks into individual songs: one set of multitrack clips per song. Chapter 5 describes how to do this under the heading "Split the Gig Recording into Song Projects." In deciding where to split the tracks, I included the audience noise before each song and the applause after each song.

The band had told me which songs to keep. The first was called "TV Mama." I saved the project as "TV Mama," deleted the tracks before and after that song's multitrack clips, and re-saved the project. In other words, I deleted all the songs before and after "TV Mama" and saved that song's multiple tracks as a separate project. (Deleting in this context does not remove any audio from the hard drive.) All the tracks for the song "TV Mama" (bass, drums, vocals, etc.) resided in a folder called "TV Mama" on my D: drive. I repeated the process for each song:

1. Load the project called "Set 1" containing the multiple tracks of all the songs.
2. Delete everything but Song 2's tracks, and save the project as "Song 2."
3. Load Set 1, delete everything but Song 3's tracks, save the project as "Song 3," and so on.

Final Mixes

Now that each song was saved as a separate project, I could start mixing. First I loaded a 16-track template into Sonar that I use for mixdowns.

Each track's fader is set to −12 dB for starters, and each track has an aux send and an EQ plug-in. Then I loaded the project for a particular song. The waveforms of all the tracks showed up on screen. *CD track 26 demonstrates the mixdown of this band's recording.*

The sax player wanted to redo some of his parts, so we brought him into the studio. He overdubbed new studio parts over the live parts. I added some short reverb to the new sax track, which made the sax sound like it was in the original venue. The keyboard player re-recorded a few solos, too.

Some of the tracks needed to be cleaned up. In the electric-guitar track, I inserted a gate plug-in to remove buzzes between musical phrases. I also gated the kick to knock out background noise between beats. The kick required some EQ to sound punchy: −6 dB at 400 Hz and +6 dB at 4 kHz. Bass, keys, and sax sounded okay as they were. I rolled some bass off the vocals to counteract the mics' proximity effect, and I compressed the vocals so that they could always be heard over the instrumental background. Most tracks needed a little reverb, set for a reverb time of 0.6 second to match the original venue.

Whenever the crowd cheered or clapped, I turned up the stereo audience track. This track was not up all the time because that can give a muddy, distant sound.

I panned each track to match the layout of the live band: sax left, bass center, vocals center, drums and keys in stereo, and lead guitar right. After tweaking the EQ and the fader automation (volume envelopes), I had a mix of the first song. I exported the song mix to a 24-bit wave file. Also, I exported the template (track configuration and settings) of that mix for use in mixing the other songs.

Finally, I loaded each song mix into Har-Bal (Harmonic Balancer), a program that automatically equalizes the spectrum of a mix to improve its tonal balance. It brings down any big peaks in the spectrum and fills in the holes, resulting in a better-sounding mix that translates well to many types of loudspeakers. I touched up some of the Har-Bal EQ. Har-Bal appended the filename of each equalized song with "EQ." Using Har-Bal is optional.

Mastering

Now all the songs' stereo mixes were on the hard drive as individual 24-bit wave files. Each one was equalized by Har-Bal. Time to master the album.

I loaded a mastering template that contained 16 tracks and a stereo output bus called Bus 1. That bus had a "maximizer" plug-in inserted, which is a peak-limiter/normalizer. It makes the CD as hot or loud as possible without affecting the dynamics. I set each track's output to Bus 1, so that all the tracks would be maximized by the same plug-in.

The band had given me the song order for the CD. I imported the first EQ'd song into track 1, imported the second EQ'd song into track 2 at the end of the first song, and so on. On screen was a series of audio clips, one per song, each on a separate track for easy time sliding and level adjustment (as in Figure 5-5).

Each song mix had some crowd noise and applause before and after each song. I overlapped the ending applause of Song 1 with the beginning of Song 2 (Figure 5-6), and so on, so that the program sounded like one continuous concert. If there wasn't enough applause for a smooth transition between songs, I copied and pasted some applause from elsewhere in the program.

With the maximizer turned off, I touched up the playback level of each track to make the songs equally loud. That was easy because all the songs were recorded and mixed at about the same level.

I noted the start time and duration of each song clip to create a cuesheet for the CdRWIN CD-burning software.

When all the edits were done, and all the songs sounded equally loud, I enabled the maximizer to see how much peak limiting could be applied without distorting the mix. I turned up the gain on the maximizer until the gain reduction reached 7 dB on the loudest peaks. More gain reduction made the music distorted. Some engineers use compression as well to make the average level louder, but this reduces dynamic range.

Then I turned on dither and exported the mastered program to a 16-bit wave file. Dither is low-level noise added to a 24-bit file just before the last 8 bits are truncated (cut off) to make a 16-bit file. This process retains most of the high resolution of the 24-bit file when it is converted to a 16-bit format for CD.

Burning the CD

Now I could prepare to burn a CD of the mastered program. I wrote a cuesheet (shown below) that listed each song's start time in the wave file. Actually, I set each start time about 1/3-second earlier, so that CD players won't miss any audio at the beginning of songs. After loading the cuesheet into the CD-burning program, I burned a test CD and listened to it. I made sure that each track started on time when I pressed the CD player's "Next track" button.

CUESHEET

```
FILE "d:\bluesband_4-23-06\bluesbandmaster.WAV" WAVE
  TRACK 01 AUDIO
    INDEX 01 00:00:00
  TRACK 02 AUDIO
    INDEX 01 10:43:26
  TRACK 03 AUDIO
    INDEX 01 18:12:10
  TRACK 04 AUDIO
    INDEX 01 26:33:12
  TRACK 05 AUDIO
    INDEX 01 30:06:21
  TRACK 06 AUDIO
    INDEX 01 37:06:21
  TRACK 07 AUDIO
    INDEX 01 40:28:12
  TRACK 08 AUDIO
    INDEX 01 46:20:22
  TRACK 09 AUDIO
    INDEX 01 54:23:04
  TRACK 10 AUDIO
    INDEX 01 59:44:03
  TRACK 11 AUDIO
    INDEX 01 64:49:04
```

I labeled the CD with a CD-marking Sharpie pen. You could use a CD-labeling kit, but a paper label can slightly increase jitter (small timing errors in the digital signal). Since this CD was a master to be duplicated, I wanted it to have as little jitter as possible. Finally, I put the CD in a soft "clamshell" case, and—after receiving payment—mailed it to the band. They were happy to get their CD: "Live at the Riverside Café."

Part 2
Classical Music Recording
(Orchestra, string quartet, pipe organ, choir, soloist)

7

MICROPHONE SPECIFICATIONS

Classical music ensembles are recorded with stereo microphone techniques. You place two or three mics in an arrangement several feet in front of the ensemble to pick up the group and the hall reverberation as a whole.

Before you can understand how stereo microphone arrays work, you need to understand microphone polar patterns. This chapter explains them, as well as other specifications that help you choose mics and accessories for stereo recording.

Polar Patterns

Microphones differ in the way they respond to sounds coming from different directions. Some mics pick up sound equally from all directions. Other mics emphasize sound from the front, but have reduced pickup of sounds coming from the sides and rear. In other words, the sensitivity of the mic varies depending on the angle of the sound source.

For example, a source in front of a mic (0° on axis) might produce a signal level from the mic that we'll call 0 dB. That same source at the side of the mic (90° off axis) might produce a signal at −6 dB. The same source at the rear of the mic (180° off axis) might produce a signal at −20 dB. A mic with this characteristic has a cardioid polar pattern. If you talk into a cardioid microphone from all sides while listening to its output, your reproduced voice will be loudest when you talk into the front of the microphone and softest when you talk into the rear.

If we graph or plot this varying sensitivity versus angle around the mic, we get a polar pattern for the microphone. Sensitivity in dB is plotted versus the angle of sound incidence in degrees. Often, several plots are made at various frequencies.

The three major polar patterns are omnidirectional, unidirectional, and bidirectional (Figure 7-1):

1. An **omnidirectional** (omni) mic is equally sensitive to sounds arriving from all directions.

2. A **unidirectional** microphone is most sensitive to sounds arriving from one direction—in front of the microphone—and rejects sounds entering the sides or rear of the microphone.

3. A **bidirectional** microphone is most sensitive to sounds arriving from two directions—in front of and behind the microphone—but rejects sounds entering the sides. Another name for "bidirectional" is "figure-eight."

Three types of unidirectional patterns are cardioid, supercardioid, and hypercardioid (Figure 7-1):

1. The **cardioid** mic is sensitive to sounds arriving from a broad angle in front of the microphone. It is about 6 dB less sensitive at the sides and about 15–25 dB less sensitive at the rear.

2. The **supercardioid** pattern is about 9 dB down at the sides and has two nulls—points of least pickup—at 125° off axis. "Off axis" means "away from the front."

3. The **hypercardioid** pattern is 12 dB down at the sides and has two nulls of least pickup at 110° either side off axis.

An omnidirectional boundary microphone (a surface-mounted microphone, explained later) has a half-omni, or hemispherical, polar pattern. A unidirectional boundary microphone has a half-supercardioid or half-cardioid polar pattern. The boundary mounting increases the directionality of the microphone and reduces pickup of room acoustics.

Some condenser mics come with switchable polar patterns.

Note that a polar plot is not a geographical map of the "reach" of a microphone; a microphone does not suddenly become dead outside its polar pattern. There is no "outside." The graph merely plots sensitivity at one frequency as distance from the origin; it is not the spatial spread of the pattern.

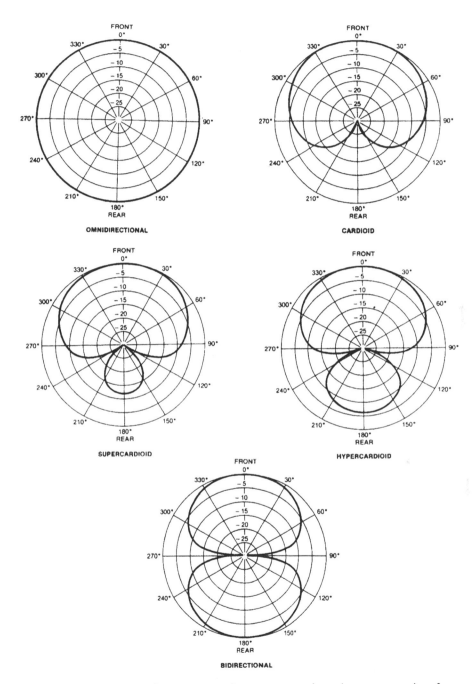

Figure 7-1 Various polar patterns. Sensitivity is plotted versus angle of sound incidence.

99

Advantages of Each Pattern

Omnidirectional microphones have several characteristics that make them especially useful for certain applications. Use omni mics when you need:

- all-around pickup,
- extra pickup of room reverberation,
- low handling noise and low wind noise,
- extended low-frequency response in omni condenser mics,
- freedom from proximity effect (up-close bass boost).

Use directional microphones when you need:

- rejection of room acoustics and background noise,
- coincident or near-coincident stereo (explained in the next chapter).

Off-Axis Coloration

In a quality microphone, the polar pattern is about the same at all frequencies between about 100 Hz and 10 kHz. If not you'll hear off-axis coloration: the mic will sound tonally different on and off axis. For example, the treble might be strong on axis and weak off axis.

Transducer Type

We've seen that microphones differ in their polar patterns. They also differ in the way they convert sound into electricity. The three operating principles of recording microphones are condenser, moving coil, and ribbon.

Of the three types, the condenser microphone generally has the widest, smoothest frequency response, and the highest sensitivity. It tends to have a natural, detailed sound, and it produces a strong signal that overrides mic-preamp noise. So it is the first choice for miking a classical music ensemble.

The condenser mic requires a power supply to operate, such as a battery or phantom power. Phantom power is provided by an external phantom-power supply, a mixer, or a mic preamp.

Ribbon mics have a smooth response, and most have a figure-eight polar pattern—making them suitable for the Blumlein stereo mic technique covered in Appendix B.

Sensitivity

Sensitivity is another specification to consider. It is a measure of the efficiency of a microphone. A very sensitive microphone produces a relatively high output voltage from a sound source of a given loudness.

Microphone sensitivity is usually stated in millivolts/Pa, where 1 Pa = 1 pascal = 94 dB SPL. The following list gives typical sensitivity specs for the three microphone types:

Condenser: 10 mV/Pa (high sensitivity)

Moving coil: 2 mV/Pa (medium sensitivity)

Ribbon or small moving coil: 1 mV/Pa (low sensitivity)

A low-sensitivity mic requires more mixer gain to achieve a good recording level than a high-sensitivity mic, and more gain usually results in more noise. If you record quiet, distant instruments, such as a classical guitar or those of a chamber music ensemble, you'll hear more mixer noise with a low-sensitivity mic than with a high-sensitivity mic, all else being equal. Because stereo miking is usually done at a distance, high sensitivity is an asset.

Self-noise

Self-noise is the electrical noise (hiss) a microphone produces. This spec is usually A-weighted—that is, the noise was measured through a filter that makes the measurement correlate more closely with the annoyance value. The filter rolls off low and high frequencies to simulate the frequency response of the human ear. An A-weighted self-noise spec of 14 dB SPL or less is excellent (quiet); a spec around 20 dB SPL is very good, and a spec around 25 dB SPL is verging on too noisy for distant miking of quiet music.

Microphone Types

Another specification is the type of microphone: the generic classification. Some types of microphones for stereo miking are free field, boundary, and stereo.

Free-Field Microphone

Most microphones are of this type. They are meant to be used in a free field—that is, away from reflective surfaces. Two types of free-field mics are the side-addressed large-diaphragm type (Figure 7-2) and the end-addressed small-diaphragm type (Figure 7-3). Both types are used to mike

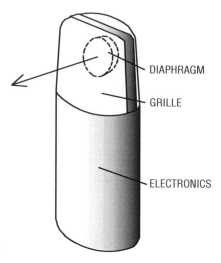

Figure 7-2 A side-addressed, large-diaphragm microphone.

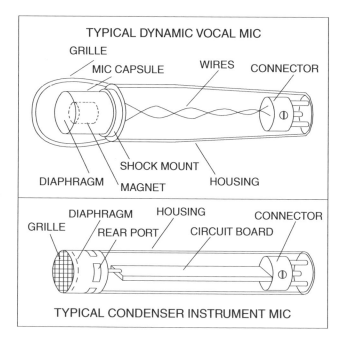

Figure 7-3 An end-addressed, small-diaphragm microphone.

orchestras. In general the large-diaphragm type has lower noise, while the small-diaphragm type has less off-axis coloration (the tone quality is almost the same on and off axis).

Boundary Microphone

A boundary microphone is designed to be used on such surfaces as a floor, wall, table, piano lid, baffle, or panel. One example of a boundary microphone is the Crown Pressure Zone Microphone (PZM) shown in Figure 7-4. It includes a miniature omni condenser capsule mounted facedown next to a sound-reflecting plate or boundary. Because of this construction, the microphone diaphragm receives direct and reflected sounds in-phase at all frequencies, avoiding phase interference between them. The claimed benefits are a wide, smooth frequency response free of phase cancellations, excellent clarity and "reach," a hemispherical polar pattern, and uniform frequency response anywhere around the microphone. Because of this last characteristic, hall reverberation is picked up without tonal coloration.

If an omnidirectional boundary mic is placed on a panel, it becomes directional. Thus, boundary mics on angled panels can be used for stereo arrays. Boundary microphones are also available with a unidirectional polar pattern. They have the benefits of both boundary mounting and the

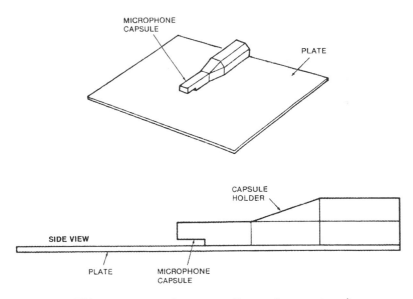

Figure 7-4 Crown PZM construction (courtesy: Crown International).

unidirectional pattern. Such microphones are well suited for stage-floor pickup of drama, musicals, or small musical ensembles.

Stereo Microphone

A stereo microphone combines two mic capsules in a single housing for convenient stereo recording. Simply place the microphone about 10–15 feet in front of a band, choir, or orchestra, and you'll get a stereo recording with little fuss. In general, a stereo microphone is easier to set up than two separate microphones, but it's more expensive.

Several models of stereo microphones are listed in Chapter 12 and one is shown in Figure 7-5.

Most stereo microphones are made with coincident microphone capsules; they occupy nearly the same point in space. Since there is no horizontal spacing between the capsules, there also is no delay or phase shift between their signals. If you combine the two stereo mic channels to mono, there is no phase interference that can degrade the frequency response. Thus, the coincident-pair stereo microphone is mono-compatible: the frequency response is the same in mono as in stereo.

Stereo microphones are available in many configurations, such as XY, MS, Blumlein, ORTF, OSS, SoundField, and SASS (all described in Appendices B and C). The MS (mid–side) stereo microphone and SoundField microphone let you remote-control the stereo spread and vary the stereo spread after recording.

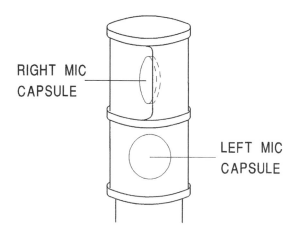

RIGHT MIC
CAPSULE

LEFT MIC
CAPSULE

Figure 7-5 A stereo microphone.

Microphone Accessories

Various accessories used with microphones enhance their convenience, aid in placement, or reduce vibration pickup.

Stands and Booms

The top of a mic stand has a standard 5/8 inch-27 thread (3/8 inch outside the US), which screws into a microphone stand adapter. Camera stores have photographic stands, which are collapsible and lightweight—ideal for recording on location. The thread is usually 1/4 inch-20, which requires an adapter to fit a 5/8 inch-27 (or 3/8 inch) thread in a mic stand adapter. Some mic stands have telescoping sections for extra height, and some have collapsible tripod stands for more stability and less weight.

You can use a mic boom as an extension to raise a microphone farther off the floor—in order to stereo mike an orchestra, for example.

Stereo Microphone Adapter

A stereo microphone adapter (stereo bar or stereo rail) mounts two microphones on a single stand for convenient stereo miking. Several models of these are listed in Chapter 12, and one is shown in Figure 7-6. In most models, the microphone spacing and angling are adjustable.

Shock Mount

When mounted on a mic stand, this device holds a microphone in a resilient suspension to isolate the microphone from mechanical vibrations, such

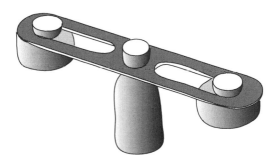

Figure 7-6 A stereo microphone adapter.

as stand and floor thumps. The shock mount acts as a spring that resonates at a sub-audible frequency with the mass of the microphone. This mass-spring system attenuates mechanical vibrations above its resonance frequency.

Many microphones have an internal shock mount that isolates the microphone capsule from its housing; this reduces handling noise as well as stand thumps.

With a good grasp of microphone specs and accessories, we're ready to discuss stereo mic techniques.

OVERVIEW OF STEREO MICROPHONE TECHNIQUES

Stereo miking is the preferred way to record classical music ensembles and soloists, such as a symphony performed in a concert hall or a string quartet piece played in a recital hall.

Stereo mic techniques capture the sound of a musical group as a whole, using only two or three microphones. When you play back a stereo recording, you hear *phantom images* of the instruments in various spots between the speakers. These image locations—left to right, front to back—correspond to the instrument locations during the recording session.

In this chapter we look at several techniques for recording in stereo.

Advantages of Stereo Miking

When recording popular music, we put a mic near each instrument, record it, and pan its image somewhere between our two monitor speakers. Then we hear where each instrument is: left, center, half-right, or whatever. But panned mono tracks are not the same as true stereo. A two-mic stereo recording captures the holistic sound of the ensemble playing together in a shared space. Large single instruments—such as piano, drums, and pipe organ—also benefit from being recorded in stereo.

Stereo miking adds lifelike realism to a recording because it captures:

- The left-to-right position of each instrument.
- The depth or distance of each instrument.
- The distance of the ensemble from the listener (the perspective).
- The spatial sense of the acoustic environment, the ambience or hall reverberation.
- The timbres of the instruments as heard in the audience.

These characteristics are lost with multiple close-up microphones.

Another advantage of stereo miking is that it tends to preserve the ensemble balance as intended by the composer. The composer has assigned dynamics (loudness notations) to the instruments in order to produce a pleasing ensemble balance in the audience area. Thus, the correct balance or mix of the ensemble occurs at a distance, where all the instruments blend together acoustically. But this balance can be upset with multiple miking. You must rely on your own judgment (and the conductor's) regarding mixer settings to produce the composer's intended balance. Of course, even a stereo pair of mics can yield a faulty balance. But a stereo pair, being at a distance, is more likely to reproduce the balance as the audience hears it.

Some outstanding examples of non-orchestral two-mic stereo recordings are those by Bob Katz (www.chesky.com), Pierre Sprey (www.mapleshaderecords.com), and Kavi Alexander (www.waterlilyacoustics.com).

Goals of Stereo Miking

One goal we aim for when miking an ensemble in stereo is accurate localization. That is, instruments in the center of the group are reproduced midway between the two speakers. Instruments at the sides of the group are heard from the left or right speaker. Instruments halfway to one side are heard halfway to one side, and so on.

Figure 8-1 shows three stereo localization effects. Figure 8-1(a) shows some instrument positions in an orchestra: left, left-center, center, right-center, right. In Figure 8-1(b), the reproduced images of these instruments are accurately localized between the speakers. The stereo spread, or stage

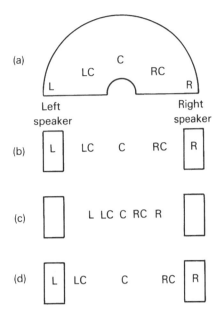

Figure 8-1 Stereo localization effects: (a) orchestra instrument locations (top view); (b) images localized accurately between speakers (the listener's perception); (c) narrow-stage effect; and (d) exaggerated separation effect.

width, extends from speaker to speaker. (You might want to record a string quartet with a narrower spread.)

A stereo pair of microphones can be angled apart, spaced apart, or both. Angling and spacing affect the stereo localization, as does the polar pattern of the microphones.

If you space or angle the mics too close together, you get a narrow stage width (Figure 8-1(c)). If you space or angle the mics too far apart, you hear exaggerated separation (Figure 8-1(d)). That is, instruments halfway to one side are heard near the left or right speaker.

To judge stereo effects, you have to sit exactly between your monitor speakers (the same distance from each). Sit as far from the speakers as the spacing between them. Then the speakers appear to be 60° apart. That is about the same angle an orchestra fills when viewed from a typical ideal seat in the audience (say, tenth-row center). If you sit off-center, the images shift toward the side on which you're sitting and are less sharp.

Play CD tracks 1–4 to set up your monitor speakers correctly for stereo listening.

Types of Stereo Mic Techniques

To make a stereo recording, you can use one of these basic techniques:

1. Coincident pair
2. Spaced pair
3. Near-coincident pair
4. Baffled pair

Let's look at each technique.

Coincident Pair

With this method (also called XY), you mount two directional mics with grilles touching, diaphragms one above the other, and angled apart (Figure 8-2). For example, mount two cardioid mics with one grille above the other, and angle them 120° apart. You can use other patterns too: super-cardioid, hypercardioid, or bidirectional. The wider the angle between mics, the wider the stereo spread.

How does this technique make images we can localize? Recall that a directional mic is most sensitive to sounds in front of the mic (on axis) and progressively less sensitive to sounds arriving off axis. That is, a directional mic puts out a high-level signal from the sound source it's aimed at, and produces lower-level signals from sources to the side of the mic.

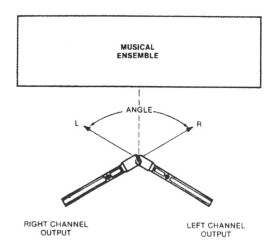

Figure 8-2 Coincident-pair technique.

The coincident pair uses two directional mics that are angled symmetrically from the center line (Figure 8-2). Instruments in the center of the group produce the same signal from each mic. When you monitor the mics, the same signal comes out of each speaker. Identical signals from two speakers produce a phantom image midway between the speakers. So you hear the center instruments in the center.

If an instrument is off-center to the right, it is more on axis to the right-aiming mic than to the left-aiming mic. So the right mic will produce a higher-level signal than the left mic. When you monitor the mics, the right speaker's signal is louder than the left speaker's signal. This reproduces the image off-center to the right. So you hear the right-side instruments toward the right side.

That is how coincident stereo miking works. The coincident pair codes instrument positions into level differences between channels. The brain decodes these level differences back into corresponding image locations. A pan pot in a mixing console works on the same principle. If one channel is 15–20 dB louder than the other, the image shifts all the way to the louder speaker. *Play CD track 5 to hear image location versus level differences between channels.*

Suppose we want the right side of the orchestra to be reproduced at the right speaker. That means the far-right musicians must produce a signal level 20 dB higher from the right mic than from the left mic. This happens when the mics are angled apart by a certain amount.

Instruments partway off-center produce interchannel level differences less than 20 dB, so you hear them partway off-center.

Listening tests have shown that coincident cardioid mics tend to reproduce the musical group with a narrow stereo spread. That is, the group does not spread all the way between speakers. *Play CD tracks 8–11 to hear the image localization of some coincident cardioid techniques.*

A coincident-pair method with excellent localization is the Blumlein array. It uses two bidirectional mics angled 90° apart and facing the left and right sides of the group.

A special form of the coincident-pair technique is the mid–side (MS) recording method illustrated in Figure 8-3. It uses a "mid" microphone facing the middle of the orchestra and a bidirectional microphone aiming to the sides. The middle mic is most commonly cardioid, but it can be any pattern.

In a device called a matrix, the signals from both mics are summed (mixed together) to produce the left-channel signal and are differenced (mixed in opposite polarity) to produce the right-channel signal.

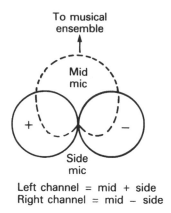

Figure 8-3 MS technique.

You can remote-control the stereo spread by changing the mid/side ratio in the matrix. This remote control is useful at live concerts, where you can't physically adjust the mics during the concert. You can also control the stereo spread during mixdown rather than during the recording. In Chapter 9 under the heading "Stereo-Spread Control," I describe how to use a computer digital audio workstation (DAW) to vary the stereo spread without using a matrix device. MS is covered in detail in Appendix B under the heading "Mid–Side."

As described in Chapter 7, a recording made with coincident mics is mono-compatible. If you expect that your recordings will be heard in mono (say, on TV), then you'll probably want to use coincident methods.

Spaced Pair

With this method (also called AB), you place two identical mics a few feet apart and aim them straight ahead (Figure 8-4). The mics can have any polar pattern, but omni is most popular for this method. The greater the spacing between mics, the greater the stereo spread.

How does this method work? Instruments in the center of the group produce the same signal from each mic. When you monitor the mics, you hear a phantom image of the center instruments midway between your speakers.

If an instrument is off-center, it is closer to one mic than the other, so its sound reaches the closer microphone before it reaches the other one. Both mics produce the same signal, except that the farther mic's signal is delayed compared to the closer mic's signal.

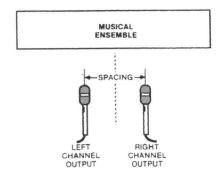

Figure 8-4 Spaced-pair technique.

If you send the same signal to two speakers with the signal in one channel delayed, the sound image shifts off-center. With a spaced-pair recording, off-center instruments produce a delay in one mic channel, so they are reproduced off-center.

The spaced pair codes instrument positions into time differences between channels. During playback, the brain decodes these time differences back into corresponding image locations. *Play CD track 6 to hear image location versus time differences between channels.*

A delay of 1.2 milliseconds (msec) is enough to shift an image all the way to one speaker. You can use this fact when you set up the mics. Suppose you want to hear the right side of the orchestra from the right speaker. The sound from the right-side musicians must reach the right mic about 1.2 milliseconds before it reaches the left mic. To make this happen, space the mics about 2–3 feet apart. This spacing makes the correct delay to place right-side instruments at the right speaker. Instruments partway off-center produce interchannel delays less than 1.2 milliseconds, so they are reproduced partway off-center.

If the spacing between mics is, say, 12 feet, then instruments that are slightly off-center produce delays between channels that are greater than 1.2 milliseconds. This places their images at the left or right speaker. I call this "exaggerated separation" or a "ping-pong" effect (Figure 8-1(d)). *Play CD tracks 18–20 to hear the image localization of some spaced-pair techniques.*

On the other hand, if the mics are too close together, the delays produced will be too small to provide much stereo spread. Also, the mics will tend to emphasize instruments in the center because the mics are closest to them.

To record a good musical balance of an orchestra, you need to space the mics about 10 or 12 feet apart. But then you get too much separation.

113

You could place a third mic midway between the outer pair and mix its output to both channels. That way, you pick up a good balance, and you hear an accurate stereo spread.

The spaced-pair method tends to make off-center images unfocused or hard to localize. Why? Spaced-pair recordings have time differences between channels. Stereo images produced solely by time differences are not very sharp. You still hear the center instruments clearly in the center, but off-center instruments are harder to pinpoint. Spaced-pair miking is a good choice if you want the sonic images to be diffuse or blended, instead of sharply focused. *Play CD track 24 to hear a comparison of a spaced pair versus a coincident pair on a drum set.*

Another flaw of spaced mics: if you mix both mic channels to mono, you may get phase cancellations of various frequencies. This may or may not be audible.

Spaced mics, however, give a "warm" sense of ambience, in which the concert-hall reverb seems to surround the instruments and, sometimes, the listener. Here's why: the two channels of recorded reverb are incoherent—that is, they have random phase relationships. Incoherent signals from stereo speakers sound diffuse and spacious. Since spaced mics pick up reverb incoherently, it sounds diffuse and spacious. The simulated spaciousness caused by the phasiness is not necessarily realistic, but it is pleasing to many listeners.

Another advantage of the spaced pair is that you can use omni mics. An omni condenser mic has deeper bass than a uni condenser mic.

Near-Coincident Pair

In this method, you angle apart two directional mics, and space their grilles a few inches apart horizontally (Figure 8-5). Even a few inches of spacing increases the stereo spread and adds a sense of ambient warmth or air to the recording. The greater the angle or spacing between mics, the greater the stereo spread.

How does this method work? Angling directional mics produces level differences between channels. Spacing mics produces time differences. The level differences and time differences combine to create the stereo effect. *Play CD track 7 to hear image location versus level and time differences between channels.*

If the angling or spacing is too great, you get exaggerated separation. If the angling or spacing is too small, you hear a narrow stereo spread.

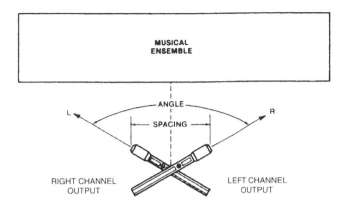

Figure 8-5 Near-coincident-pair technique.

A common near-coincident method is the ORTF (French) system, which uses two cardioids angled 110° apart and spaced 7 inches (17 cm) horizontally. Usually this method gives accurate localization. That is, instruments at the sides of the orchestra are reproduced at or very near the speakers, and instruments halfway to one side are reproduced about halfway to one side.

The NOS (Dutch) system uses two cardioids angled 90° and spaced 12 inches (30 cm), while the DIN (German) system is 90° and 7.9 inches (20 cm). Compared to ORTF, those methods have less off-axis coloration because the mics are less angled away from the center instruments. Also their 90° angle between mics is easier to set up visually than the ORTF 110° angle between mics. *Play CD tracks 12–14 to hear the image localization of some near-coincident-pair methods.*

Baffled-Omni Pair

This method uses two omni mics, usually ear-spaced, and separated by a hard or padded baffle (Figure 8-6). To create stereo, it uses time differences at low frequencies and level differences at high frequencies. The spacing between mics creates time differences. The baffle creates a sound shadow (reduced high frequencies) at the mic farthest from the source. Between the two channels, there are spectral differences (differences in frequency response).

115

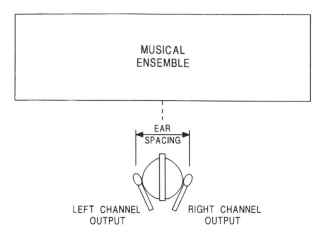

Figure 8-6 Baffled-omni technique.

Figure 8-7 Sphere microphone.

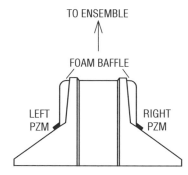

Figure 8-8 Crown SASS-P MKII stereo PZM (Pressure Zone Microphone) microphone (top view).

Some examples of baffled-omni pairs are the Schoeps or Neumann sphere microphones (Figure 8-7), the Jecklin Disk, and the Crown SASS-P MKII stereo microphone (Figure 8-8). The omni condenser mics used in the baffled-omni method have excellent low-frequency response. *Play CD tracks 15–17 to hear the image localization of some baffled-omni-pair methods.*

A special form of the baffled-omni pair is binaural recording with an artificial head (dummy head). The head contains a microphone flush mounted in each ear. You record with these microphones and play back the recording over headphones. This process can re-create the locations of the original performers and their acoustic environment with startling realism. For more detail on binaural recording see Appendix D.

You can clip a pair of miniature omni or cardioid mics onto the temple pieces of eyeglasses. Each mic is on the opposite side of your head, either in your ears or on your temples. Chapter 12 lists some manufacturers of these binaural microphones. To compensate for the acoustic effect of the head, the signals need some equalization (EQ) (a broad dip around 3 kHz).

Boundary (surface-mounted) mics can be used for any type of stereo miking. Appendix C describes some stereo mic techniques using boundary mics. *Play CD tracks 21–23 to hear the image localization of some stereo boundary-mic techniques.*

Comparing the Four Techniques

1. Coincident pair
   - Uses two directional mics angled apart with grilles touching.
   - Level differences between channels produce the stereo effect.
   - Images are sharp.
   - Stereo spread ranges from narrow to accurate.
   - Signals are mono-compatible.
2. Spaced pair
   - Uses two mics spaced a few feet apart, aiming straight ahead.
   - Time differences between channels produce the stereo effect.
   - Off-center images are diffuse.
   - Stereo spread tends to be exaggerated unless a third center mic is used, or unless spacing is under 2–3 feet.
   - Provides a warm sense of ambience.
   - Provides excellent low-frequency response if you use omni condensers.
   - Tends not to be mono-compatible, but this might not be audible.

3. Near-coincident pair
   - Uses two directional mics angled apart and spaced a few inches apart horizontally.
   - Level and time differences between channels produce the stereo effect.
   - Images are sharp.
   - Stereo spread tends to be accurate.
   - The hall sounds more spacious than with coincident methods.
   - Tends not to be mono-compatible.
4. Baffled-omni pair
   - Uses two omni mics, usually ear-spaced, with a baffle between them.
   - Level, time, and spectral differences produce the stereo effect.
   - Images are sharp.
   - Stereo spread tends to be accurate.
   - Excellent low-frequency response.
   - Good imaging with headphones.
   - The hall sounds more spacious than with coincident methods.
   - Stereo spread is not adjustable except by panning the two channels toward the center.
   - More conspicuous than other methods.
   - Tends not to be mono-compatible, but this might not be audible.

Mic Requirements for Stereo

For sharp imaging, the microphone pair should be well matched in frequency response and polar pattern. Be sure both mics are the same model number, and match their levels when picking up a sound source in the center. Or use a stereo mic, which mounts two mic capsules in a single housing for convenience.

Surprisingly, different transducer types have different imaging. Why? For sharpest imaging, microphone polar patterns and off-axis phase shift should be uniform with frequency. In a ribbon mic, these needs are met. But a condenser mic tends to be less uniform with frequency, and a dynamic tends to be still less uniform. These characteristics affect the imaging of a stereo pair of microphones.

How to Test Imaging

Here is a way to check the stereo imaging of a mic technique:

1. Set up the stereo mic array in front of a stage.
2. Record yourself speaking from various locations on stage where the instruments will be: center, half-right, far right, half-left, far left. Announce your position.
3. Play back the recording over speakers. Sit exactly between them, as far away from them as they are spaced apart.

You'll hear how accurately the technique translated your positions, and you'll hear how sharp the images are. If you hear a narrow stereo spread, you need to angle or space the mics farther apart. If you hear exaggerated separation, you need to angle or space the mics closer together.

We looked at several mic arrays to record in stereo. Each has its pros and cons. Which method you choose depends on the sonic compromises you're willing to make.

Recommended Reading

Blumlein, A. "British Patent Specification 394,325." *Journal of Audio Engineering Society*, Vol. 6, No. 2 (April 1958), p. 91.

Keller, A. "Early Hi Fi and Stereo Recording at Bell Laboratories (1931–1932)." *Journal of Audio Engineering Society*, Vol. 29, No. 4 (April 1981), pp. 274–280.

These references can be found in *Stereophonic Techniques*, an anthology published by the Audio Engineering Society, 60 E. 42nd Street, New York, NY, 10165.

9

STEREO RECORDING PROCEDURES

This chapter explains how to do an on-location stereo recording of a classical music ensemble. We cover both equipment and procedures.

Equipment

- Microphones (low-noise condenser or ribbon type, omni or directional, free field or boundary, stereo or matched pair).
- Multitrack recorder or stereo recorder. These were described in Chapter 1 in the sections "Multitrack Recorder" and "Stereo Recorder."
- Low-noise mic preamps or low-noise mini mixer (necessary if the mic preamps in your recorder are low quality).
- Phantom-power supply (necessary if your mixer or mic preamp lacks phantom power).
- Stereo bar.
- Mic stands and booms, or fishing line to hang mics.
- Shock mount (optional).
- Mixer (necessary if you use more than two mics).
- Mid–side (MS) matrix box (if you are recording with the MS technique).

- Headphones and/or speakers.
- Power amplifier for speakers (optional) or powered Nearfield monitors.
- Recording medium: hard drive or flash memory.
- Power strip and extension cords.
- Notebook and pen.
- Talkback mic and powered speaker (optional).
- Tool kit.
- Fresh batteries.

First on the list are microphones. You'll need at least two or three of the same model number or one or two stereo microphones. Stereo and surround microphones are listed in Chapter 12 under the headings "Stereo Microphones" and "Surround Microphones." Good microphones are essential, for the microphones—and their placement—determine the sound of your recording. To achieve professional-quality recordings of music, you should expect to spend at least $100 per microphone or $1000 for a stereo microphone.

For classical music recording, the preferred microphones are condenser or ribbon types with a wide, flat frequency response and very low self-noise (explained in Chapter 7). A self-noise spec of less than 20 dB equivalent SPL, A-weighted, is recommended.

If you want to do spaced-pair recording, you can use either omnidirectional or directional microphones. Omnis are preferred because they generally have a deeper low-frequency response. If you want to do coincident or near-coincident recording for sharper imaging, use directional microphones (cardioid, supercardioid, hypercardioid, or bidirectional). Mics with those polar patterns tend to roll off in the low frequencies, but you can compensate for this with equalization.

You need a power supply for condenser microphones: either an external phantom-power supply, a mixer or mic preamp with phantom power, or internal batteries. A low-noise stereo mic preamp (or low-noise portable mixer) lets you make recordings free of electronic hiss.

You can mount the microphones on stands or hang them from the ceiling with nylon fishing line. Make sure that the fishing line's tensile strength exceeds the weight of the mics. Check legal safety issues with hanging mics; different rules apply in different places. Stands are much easier to set up, but more visually distracting at live concerts. Stands are more suitable for recording rehearsals or sessions with no audience present.

Neumann makes tiltable "auditorium hangers" that suspend mic capsules of the KM 100 system from their cables.

The mic stands should have a tripod-folding base and should extend at least 13 feet high. Some suitable products are telescoping photographic stands (available from camera stores such as www.prostudiousa.com). They are lightweight and compact. Other examples are the Shure S15A (www.shure.com), Quik Lok A85 (www.quiklok.com), AEA-13MDV (www.wesdooley.com), K&M 21411B, and various models at www.micsupply.com. You can use baby booms or stand extenders to increase the height of regular mic stands.

A useful accessory is a stereo bar (stereo microphone adapter, stereo mic mount). This device mounts two microphones on a single stand for coincident or near-coincident stereo recording. A long stereo bar can accommodate spaced-pair miking. Some examples are given in Chapter 12 under the heading "Stereo and Surround Microphone Adapters."

Another needed accessory in most cases is a shock mount to prevent pickup of floor vibrations. Some models are the On-Stage MY410 (various vendors), Sabra Som SSM-1 (www.sabra-som.com), Shure A55M (www.shure.com), and AKG H85 and H100 (www.akg.com).

In difficult mounting situations, boundary microphones may come in handy. They can lie flat on the stage floor to pick up small ensembles or can be mounted on the ceiling or on the front edge of a balcony. They also can be attached to clear Plexiglas panels that are hung or mounted on mic stands. Boundary mics are made by most microphone companies.

To monitor in the same room as the musicians, you need some closed-cup, circumaural (around the ear) headphones, or isolating earphones to block out the sound of the musicians. You want to hear only what is being recorded. Of course, the headphones should be wide-range and smooth for accurate monitoring. A better monitoring arrangement might be to set up powered Nearfield loudspeakers in a separate room.

If you're in the same room as the musicians, you have to sit far from them to clearly monitor what you're recording. To do that, you need a pair of 50-foot microphone extension cables. Longer extensions are needed if the mics are hung from the ceiling or if you monitor in a separate room.

A mixer is necessary when you want to record more than one source—for example, an orchestra and a choir, an orchestra with spot mics, or a band and a soloist. You might put a pair of microphones on the orchestra and another pair on the choir. Connect the insert sends from the mixer to a multitrack recorder, then mix the tracks to stereo back in the studio. If your budget doesn't allow a multitrack recorder you could mix the mic signals live to a stereo recorder.

For monitoring an MS recording, bring an MS matrix box that converts the MS signals to left-right signals, which you monitor. MS matrix devices are listed in Chapter 12 under the heading "MS Matrix Decoders."

Selecting a Venue

If possible, plan to record in a venue with good acoustics. It should have adequate reverberation time for the music being performed (about 2 seconds for orchestral music). This is very important, because it can make the difference between an amateur-sounding recording and a commercial-sounding one. Try to record in an auditorium, concert hall, or house of worship rather than in a band room or gymnasium. Halls with wooden surfaces and a shoebox shape tend to sound best.

You may be forced to record in a hall that is too dead: that is, the reverberation time is too short. In this case, you may want to add artificial reverberation from a digital reverb unit or plug-in, or cover the seats with plywood sheets or 4-mm polyethylene plastic sheeting. Strong echoes can be controlled with carpets, RPG diffusors, or drapes.

If the venue is surrounded by noisy traffic, consider recording after midnight. Turn off noisy air conditioning if possible while recording.

Call the venue manager and ask that the circuit breakers for the stage outlets be turned on the day of the session. Ask where you can load-in your equipment. Also make sure that the load-in door will be unlocked when you plan to load in.

Session Setup

Be sure to test all your equipment for correct operation before going on the job. If you are using battery-powered devices, install fresh batteries in them just before the concert. Keep your equipment inside your home or studio until you're ready to leave. A recorder left outside in a cold car may become sluggish if the lubricant stiffens, and batteries may lose some voltage.

Make sure your hard drive or flash card has enough capacity for the session. Table 3-1 in Chapter 3 shows the storage required for a 1-hour recording.

Arrive at the venue a few hours early to allow for setup and for fixing problems.

First, power-up the equipment. You can use batteries or an AC extension cord plugged into an outlet near the stage. Make sure this outlet is live. Gaffer-tape the extension cord lengthwise or cover it with mats to prevent tripping. Tie your outlet strip's AC cord to the extension cord, so that they can't separate if someone pulls on the extension cord.

If this is a session, listen to the ensemble playing on the stage. Is the sound bad? You might want to move the musicians out onto the floor of the hall.

Mounting the Mics

Place your microphones in the desired stereo miking arrangement. As an example, suppose you're recording an orchestra rehearsal with two crossed cardioids on a stereo bar (the near-coincident method). Screw the stereo bar onto a mic stand and mount two cardioid microphones on the stereo bar. For starters, angle them 110° apart and space them 7 inches apart horizontally. Aim them downward, so that they'll point at the orchestra when raised. You may want to mount the microphones in shock mounts or put the stands on sponges to isolate the mics from floor vibration.

Basically, place two or three mics (or a stereo mic) several feet in front of the group, raised up high (as in Figure 9-1). The microphone placement controls the acoustic perspective or sense of distance to the ensemble, the balance among instruments, and the stereo imaging.

As a starting position, place the mic stand behind the conductor's podium, about 12 feet from the front-row musicians. Connect mic cables and gaffer-tape them to the top of each mic stand for strain relief. Raise the microphones about 14 feet off the floor. This prevents overly loud pickup of the front row relative to the back row of the orchestra. It also reduces audience noise by mounting the mics farther from the audience. Gaffer-tape the mic cables to the bottom of the stand to keep it from being pulled over.

Leave some extra turns of mic cable at the base of each stand, so that you can reposition the stands. This slack also allows for people accidentally pulling on the cables. Try to route the mic cables where they won't be stepped on, or cover them with mats.

Live, broadcast, or filmed concerts require an inconspicuous mic placement, which may not be sonically ideal. In these cases, or for permanent installations, you probably want to hang the microphones from the ceiling rather than using stands. You can position a stereo bar with

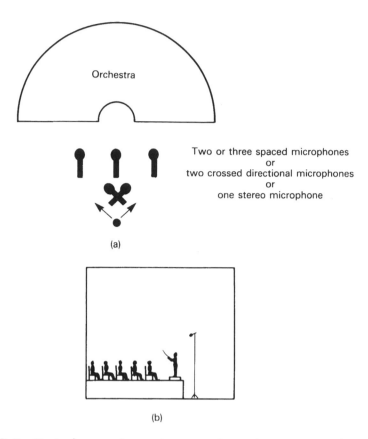

Figure 9-1 Typical microphone placement for on-location recording of a classical music ensemble: (a) top view and (b) side view.

three nylon fishing lines spaced apart. Make sure that the tensile strength of the fishing line exceeds the weight of the mics and stereo bar. Hang individual mics by their cables and attach two fishing lines to the front of each mic to aim it. Another inconspicuous mic placement is on mic-stand booms projecting forward of a balcony. For dramas or musicals, directional boundary mics can be placed on the stage floor near the footlights.

If you are uncertain which mic technique to use, you could set up two or more mic arrays and record them to multitrack. When you master the program, choose the best-sounding technique. You might even mix the signals of two or more mic pairs.

Connections

Now you're ready to make connections. Here are some connection methods for just two mics:

- If your stereo recorder has high-quality mic preamps and phantom power, plug the mics directly into the recorder mic inputs.
- If your recorder lacks phantom power (or the phantom voltage is lower than the mics require), plug the mics into a phantom supply, and from there into your recorder mic inputs.
- If your recorder lacks high-quality mic preamps, plug the mics into a low-noise mic preamp or mixer with phantom power. Connect cables from there into your recorder line inputs.

Here are some connection methods for more than two mics:

- Plug the mics into a stage box (described in Chapter 1) and run the snake back to your mixer. Plug the snake mic connectors into your mixer mic inputs. If you are recording to two-track, plug the mixer stereo outputs into the recorder line inputs. If you are recording to multitrack, plug the mixer insert-send jacks into the recorder line inputs.
- If you have a multichannel audio interface with mic preamps, plug the mics into the interface. Connect the interface USB or FireWire port to a laptop computer.
- If you want to feed your mic signals to several mixers—recording, broadcast, PA—plug your mic cables into a mic splitter or distribution amp (described in Chapter 2 under the heading "Splitting the Microphones"). Connect the splitter outputs to the snakes for each mixer. Supply phantom from one mixer only, on the direct side of the split. Each split will have a ground-lift switch on the splitter. Set it to *ground* for the mixer supplying phantom. Set it to *lift* for the other mixers (or to *ground* if that results in the least hum). This prevents hum caused by ground loops between the different mixers.
- If you're using directional microphones and want to make their response flat at low frequencies, you can run them through a mixer with equalization for bass boost. You might prefer to equalize the recording during mastering instead. Boost around 50–100 Hz until the bass sounds natural or until it matches the bass response of omni condenser mics. You won't need this equalization (EQ) if the microphones have been factory equalized for flat response at a distance.

Monitoring

Put on your headphones or listen over loudspeakers in a separate room. Sit equidistant from the speakers, as far from them as they are spaced apart. You'll probably need to use a Nearfield arrangement (speakers about 3 feet apart and 3 feet from you) to reduce coloration of the speakers' sound from the room acoustics. *Play CD tracks 1–4 to set up your monitor speakers correctly for stereo listening.*

Turn up the recording-level controls and monitor the signal. When the orchestra starts to tune up, set the recording levels to peak around −15 dB, so that you have a clean signal to monitor. You can set levels more carefully later on.

Microphone Placement

Nothing has more effect on the production style of a classical music recording than microphone placement. Miking distance, polar patterns, angling, spacing, and spot miking all influence the recorded sound's character.

Miking Distance

The microphones must be placed closer to the musicians than a good live listening position. If you place the mics out in the audience where the live sound is good, the recording probably will sound muddy and distant when played over speakers. That is because all the recorded reverberation is reproduced up-front along a line between the monitor speakers, along with the direct sound of the orchestra. Close miking (5–20 feet from the front row) compensates for this effect by increasing the ratio of direct sound to reverberant sound.

The closer the mics are to the orchestra, the closer the orchestra sounds in the recording. If the instruments sound too close, too edgy, or too detailed, or if the recording lacks hall ambience, the mics are too close to the ensemble. Move the mic stand a foot or two farther from the orchestra and listen again.

If the orchestra sounds too distant, muddy, or reverberant, the mics are too far from the ensemble. Move the mic stand a little closer to the musicians and listen again.

Eventually, you'll find a sweet spot, where the direct sound of the orchestra is in a pleasing balance with the ambience of the concert hall. Then the reproduced orchestra will sound neither too close nor too far.

Here's why miking distance affects the perceived closeness (perspective) of the musical ensemble: the level of reverberation is fairly constant throughout a room, but the level of the direct sound from the ensemble increases as you get closer to it. Close miking picks up a high ratio of direct-to-reverberant sound; distant miking picks up a low ratio. The higher the direct-to-reverb ratio, the closer the sound source is perceived to be. *Play CD track 25 to hear how miking distance affects the direct-to-reverb ratio.*

If the recording venue is "live" because of hard surfaces—brick, glass, stone—chances are you will need to mike closely. On the other hand, if the venue is "dead" because of soft surfaces—carpet, drapes, stuffed seats—expect to mike farther away.

An alternative to finding the sweet spot is to place a stereo pair close to the ensemble (for clarity) and another stereo pair distant from the ensemble (for ambience). According to Delos recording director John Eargle, the distant pair should be no more than 30 feet from the main pair, otherwise the signal might simulate an echo. You record the two pairs to a multitrack recorder and mix them back in the studio. The advantages of this method are as follows:

- It avoids pickup of bad-sounding early reflections.
- Close miking reduces pickup of background noise.
- After the recording is finished, you can adjust the direct/reverb ratio or the perceived distance to the ensemble.
- Comb filtering due to phase cancellations between the two pairs is not severe because the delay between them is great and their levels and spectra are different.

Similarly, Skip Pizzi recommends a "double MS" technique, which uses a close MS microphone mixed with a distant MS microphone (as shown in Figure 9-2). One MS microphone is close to the performing ensemble for clarity and sharp imaging, and the other is out in the hall for ambience and depth. The distant mic could be replaced by an XY pair for lower cost.

If the ensemble is being amplified through a sound-reinforcement system, you might be forced to mike very close to avoid picking up amplified sound and feedback from the reinforcement speakers. In that case you will need to add high-quality artificial reverberation or convolution reverb.

For broadcast or communications, consider miking the conductor with a wireless lavalier mic or stand-mounted mic.

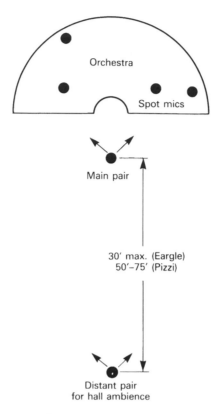

Figure 9-2 Double MS technique using a close main pair and a distant pair for ambience. Spot mics also are shown.

Stereo-Spread Control

Concentrate on the stereo spread. If the monitored spread is too narrow, it means that the mics are angled or spaced too close together. Increase the angle or spacing between mics until localization is accurate. *Note*: Increasing the angle between mics will make the instruments sound farther away; increasing the spacing will not. *Play CD track 9. The mics are angled and spaced too close together, so the stereo spread is narrow.*

If off-center instruments are heard far left or far right, that indicates your mics are angled or spaced too far apart. Move them closer together until localization is accurate. *Play CD track 20. The mics are too far apart, so the stereo separation is exaggerated.*

If you record with an MS microphone, you can change the monitored stereo spread either during the recording or after.

To change the spread during the recording, connect the stereo mic outputs to the matrix box and connect the matrix-box L—R output to the recorder. Use the stereo-spread control (*M/S* ratio) in the matrix box to adjust the stereo spread.

To alter the spread after the recording using a matrix box: Record the mid signal on one track and the side signal on another track. Monitor the output of the recorder with a matrix box. Back in the studio, run the mid and side tracks through the matrix box, adjust the stereo spread as desired, and record the left and right outputs.

To alter the spread after the recording using a digital audio workstation (DAW):

1. Record the mid mic on track 1; record the side mic on track 2.
2. Copy or clone track 2 to track 3. Be sure the waveforms are aligned.
3. Pan track 2 hard left; pan track 3 hard right.
4. Reverse the polarity of track 3 or use an "invert polarity" plug-in.
5. Group tracks 2 and 3, so their faders move together.
6. To change the stereo spread, vary the levels of tracks 2 and 3 relative to track 1.

If you are set up before the musicians arrive, check the localization by recording yourself speaking from various positions in front of the microphone pair while announcing your position ("left side," "mid-left," "center"). Play back the recording to judge the localization accuracy provided by your chosen stereo array. Recording this localization test is an excellent practice. *Play CD tracks 8–24 to hear the localization of several stereo mic techniques.*

Monitoring Stereo Spread

Full stereo spread on speakers is a spread of images all the way between speakers, from the left speaker to the right speaker. Full stereo spread on headphones can be described as stereo spread from ear to ear. The stereo spread heard on headphones may or may not match the stereo spread heard over speakers, depending on the microphone technique used.

Due to psychoacoustic phenomena, coincident-pair recordings have less stereo spread over headphones than over loudspeakers. Take this into account when monitoring with headphones or use only loudspeakers for monitoring. *Play CD tracks 9–10 alternately over loudspeakers and headphones. Compare the stereo spread.*

If you are monitoring your recording over headphones or anticipate headphone listening to the playback, you may want to use near-coincident miking techniques, which have similar stereo spread on headphones and loudspeakers. *Play CD tracks 13–14 alternately over loudspeakers and headphones. Compare the stereo spread.*

Ideally, monitor speakers should be set up in a Nearfield arrangement (say, 3 feet from you and 3 feet apart) to reduce the influence of room acoustics and to improve stereo imaging. On the wall behind the monitors, attach a panel of acoustic foam that extends a few feet beyond the speaker spacing.

If you want to use large monitor speakers placed farther away from you, deaden the control-room acoustics with acoustic foam or thick fiberglass insulation (covered with muslin). Place the acoustic treatment on the walls behind and to the sides of the loudspeakers. This smoothes the frequency response and sharpens stereo imaging.

You might include a stereo/mono switch in your monitoring system to check for mono compatibility.

Soloist Pickup and Spot Microphones

Sometimes a soloist plays in front of the orchestra. You have to capture a tasteful balance between the soloist and the ensemble. That is, the main stereo pair should be placed so that the relative loudness of the soloist and the accompaniment is musically appropriate. If the soloist is too loud relative to the orchestra (as heard on headphones or loudspeakers), raise the mics. If the soloist is too quiet, lower the mics. You may want to add a spot mic (accent mic) about 3 feet from the soloist and mix it with the other microphones. Take care that the soloist appears at the proper depth relative to the orchestra.

If a sound-reinforcement system is in use, place the soloist mic 8–12 inches away to prevent feedback. Use a foam windscreen on a vocal mic. To make the soloist mic inconspicuous, you could place a small-diaphragm mic at about chest height aiming at the mouth, and use a slender mic stand such as the Schoeps Active Tube RC.

Many record companies prefer to use multiple microphones and multitrack techniques when recording classical music. Such methods provide extra control of balance and definition and are necessary in difficult situations. Often you must add spot or accent mics on various instruments or instrumental sections to improve the balance or enhance clarity (as shown in Figure 9-2). In fact, John Eargle contends that a single stereo pair of mics rarely works well.

A choir that sings behind the orchestra can be miked separately with two to four cardioids. You might place the choir in the audience area facing the orchestra, and mike the choir.

If the recording mics are also used for sound reinforcement, place the choir mics about 3 feet out front and 3 feet above the head height of the back row. Add piano mics and windscreened soloist mics about 8 inches from their sound sources. Record all the mics to multitrack, and mix the tracks back in your studio. Add reverb and EQ as needed.

Pan each spot mic so that its image position coincides with that of the main microphone pair. Using the mute switches on your mixing console, alternately monitor the main pair and each spot mic to compare image positions.

You might want to use an MS microphone or stereo pair for each spot mic. Adjust the stereo spread of each local sound source to match that reproduced by the main pair. For example, suppose that a violin section appears 20° wide as picked up by the main pair. Adjust the perceived stereo spread of the MS spot mic used on the violin section to 20°, then pan the center of the section image to the same position that it appears with the main mic pair.

When you use spot mics, mix them at a low level relative to the main pair—just loud enough to add definition but not loud enough to destroy depth. Operate the spot-mic faders subtly or leave them untouched. Otherwise the close-miked instruments may seem to jump forward when the fader is brought up, then fall back in when the fader is brought down. If you bring up a spot-mic fader for a solo, drop it only 6 dB when the solo is over, not all the way off.

Often the timbre of the instrument(s) picked up by the spot mic is too bright. You can fix it with a high-frequency rolloff or by using a mic with that characteristic. Adding artificial reverb to the spot mic can help too.

To further integrate the sound of the spots with the main pair, you might want to delay each spot's signal to coincide with those of the main pair. That way, the main and spot signals are heard at the same time. For each spot mic, the formula for the required delay is:

$$T = D/C$$

where
T: delay time in seconds
D: distance between each spot mic and the main pair in feet
C: speed of sound, 1130 feet/second.

133

For example, if a spot mic is 20 feet in front of the main pair, the required delay is 20/1130 or 17.7 milliseconds. Some engineers add even more delay (10–15 milliseconds) to the spot mics to make them less noticeable (Streicher and Dooley). As a rule of thumb, 1 foot corresponds to about 1 millisecond of delay.

A suggested track assignment for multitrack concert recordings is shown below:

Tracks 1–2: main pair

Tracks 3–4: distant pair

Tracks 5–6: "outriggers" or widely spaced pair

Tracks 7–8: spot mics and conductor's mic (for announcing takes).

Once the microphones are positioned properly, gaffer-tape the mic-stand legs to the floor, so that the stands can't be knocked over.

Setting Levels

Now you're ready to set recording levels. Ask the conductor to have the orchestra play the loudest part of the composition, and set the recording level for the desired meter reading. A typical recording level is −6 dB maximum on a peak-reading meter for a digital recorder. The level can go up to 0 dB maximum without distortion, but aiming for −6 dB allows for surprises. Bass-drum and tympani hits produce the highest peaks.

If you plan to record a concert with no sound check, you have to set the record-level knobs to a nearly correct position ahead of time. Do this during a pre-concert trial recording, or just go by experience: set the knobs where you did at previous sessions (assuming you are using the same microphones in the same venue). Another way to pre-set the recording level is to aim for a peak meter level of −15 dB when the orchestra tunes up.

Before going on-location, you could play loud orchestral music over your studio monitors or home stereo, set up your mics and recorder, and set approximate recording levels. If you need to set the level less than one-third up to get a 0-dB meter level, you probably need to insert a pad or set the mic-gain switch to low gain.

Recording a Concert

Before the concert obtain a printed program of the musical selections. On this program, next to each piece, you will write down the recorder counter times of the beginning and end of that piece. That will help you locate and identify the pieces correctly when you edit the recording later.

Start recording when the conductor walks out (or sooner). Record the concert nonstop if your recording medium allows. Document your mic placement and recording level for future jobs.

Editing

At this point, the recording is finished and you've brought it back to your studio. The concert recording has long applause after each piece. Suppose you want to edit the applause shorter and insert a few seconds of silence between the compositions. Here is a suggested procedure for editing a stereo recording using a DAW:

1. Connect a USB cable between the computer and recorder. Then the computer recognizes the recorder as a storage device. Click-drag the recorder's wave file of the concert to your hard drive. If you recorded to DAT, copy the program in real time through your sound card or audio interface (ideally in digital format).
2. In the audio editing software, start a new project and set up an audio track.
3. Import the concert file from the hard drive into the audio track.
4. Play the track and locate the first piece. Refer to the counter times you wrote on your session notes or concert program.
5. Delete the part of the recording before the first piece but do not close up the space.
6. Find the applause at the end of the piece. About 10 seconds into the applause, split the clip or region. Also split the clip a few seconds before the start of the next piece. Cut out the audio between the split points, but do not close up the space.
7. Label the clip of the composition with its title.
8. Repeat Step 6 for the rest of the pieces.

Now each musical piece is in a separate clip or region on your screen. Slide the first piece to the beginning of the program. Next you will add fades and adjust the spacing between the pieces (refer to Figure 9-3):

1. At the end of each piece, let the applause play for 3 seconds then fade it out over about 8 seconds Use a fade that starts quickly and ends slowly.

135

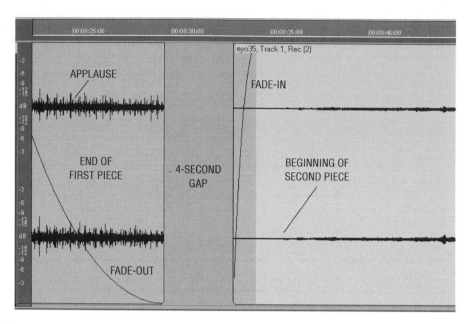

Figure 9-3 A DAW screen showing an applause fade-out, a 4-second gap, and a fade-in before the next piece.

2. If there is background noise such as air-conditioning rumble, insert a fade-in about 2 seconds before the beginning of each piece (Figure 9-3). Or you may want the track to start right when the music starts; that is, with no ambience before it starts.

3. Time-slide the clips to create a 4-second gap between them (or whatever interval sounds right).

If you don't want total silence between pieces, do not fade out the applause. End each piece's clip where the applause stops, and leave about 4 seconds of recorded "room tone" between clips.

To edit a session that has no applause, set the beginning of each clip just before the first note. Set the end of each clip just after the reverberant tail fades to silence. Leave about a 4-second silent gap between clips (or whatever sounds right).

When you're finished editing, note the start time and duration of each clip. You will use this information to write a cuesheet that determines the start IDs when burning a CD. Export the edited program to a 24-bit stereo file.

Sometimes, despite your best intentions and the finest microphones, the spectral balance of the recording might be poor. The bass might be weak, the strings strident, and so on. Room acoustics play a strong part in this. Fortunately, a recording having a skewed tonal balance can often be salvaged with equalization. An effective tool for this purpose is Harmonic Balancer (www.har-bal.com), which shows the spectrum of the recording and lets you equalize it as needed.

Once the program is equalized (if necessary), import the equalized file and normalize it, so that the highest peak reaches −0.1 dBFS. Finally, enable dither and export the normalized mix to a 16-bit stereo file. You will burn a CD from this file.

A Real-World Example: Recording an Orchestra in a Concert Hall

To illustrate what an actual concert recording might be like, I'll describe an on-location project that I did recently. The promoter of a local symphony orchestra phoned and asked me to record the group live in concert. She wanted 25 CDs of the concert for the musicians.

Preproduction

In talking with the promoter, I got this information:

Performance date and time: May 9, 2006, at 7 p.m.

Load-in time: 6 p.m.

Length of performance: 1 hour, no intermission.

Venue and venue address: Venerable Theater, 110 S. River St., Elkhart, IN.

Instrumentation: Symphonic orchestra with no soloists.

Directions to venue: I used Mapquest (www.mapquest.com).

Next I drew a block diagram of the recording system. I planned to record the orchestra with a pair of mics and a flash-memory recorder. From the block diagram, I generated an equipment list:

2 Neumann KM140 cardioid mics

Stereo mic adapter

Tall foldable mic stand and boom extender

Two 15-foot mic cables and a spare mic cable

ART Phantom II battery-powered phantom power supply

Behringer UBB 1002 battery-powered mixer

Edirol R-09 battery-powered flash-memory recorder

AKG K240 headphones

Spare batteries

Carrying case

Gaffers tape

Pen, notebook, and flashlight

All this fit into a cloth bag and a carrying case.

In the studio, I set up all this equipment to test it. I screwed the stereo mic adapter onto the boom extender, and taped the boom extender temporarily to the folded mic stand to make carrying easier.

The Recording Session

I arrived at the session an hour before showtime. Setup for this gig typically takes 20 minutes.

I have recorded at this venue before and have found that the best spot for the microphones is in the 3rd row center, raised about 14 feet. That places the mics about 12 feet behind the conductor. In this particular venue, placement closer than 12 feet gives a dry, bright sound and overemphasizes the front-row strings. With the mics farther than 12 feet, the recorded orchestra sounds too distant and less engaging.

After unpacking, I mounted the two Neumann KM140 cardioid mics in the stereo mic adapter in the NOS configuration, aiming down to where the orchestra would be when the mics are raised. After plugging the cables into the mics, I formed a sideways "U" out of the mic cables and taped the cables to the boom extension to act as a strain relief. Next, I screwed the boom extension onto the mic stand and raised the mics all the way.

The mixer, phantom supply, and recorder were in a carrying case in my lap. I plugged the mic cables into the phantom supply, which was plugged into the mixer and from there into the recorder line input. Why were those devices necessary? The recorder mic input was a little noisy and did not provide 48V phantom, which the Neumann mics required. The battery-powered mixer had quiet mic preamps, but it also did not provide 48V phantom, so I needed an external 48V phantom supply.

Once everything was connected, I powered up the recording devices and listened to the mic signals over headphones. I verified that the left and right channels were not reversed. When the orchestra tuned up I set the recording level to peak at -15 dB.

Finally, the conductor walked onstage. I started recording and noted the counter time of each piece. The concert was recorded nonstop for editing later.

Editing/Mastering

Back in the studio, I set up my DAW to edit the concert. I followed the procedure given in this chapter under the heading "Editing." I burned a master CD and a safety copy. Finally, I duplicated the master CD several times, put on CD labels, and delivered a box of CDs to the conductor.

References

Pizzi, S. "Stereo Microphone Techniques for Broadcast." *Audio Engineering Society Preprint No. 2146 (D-3), Presented at the 76th Convention*, October 8–11, 1984, New York.

Streicher, R., and Dooley, W. "Basic Stereo Microphone Perspectives—A Review." *Journal of the Audio Engineering Society*, Vol. 33, No. 7/8 (July/August 1985), pp. 548–556.

Eargle, J. *The Microphone Book*. Boston: Focal Press, 2001.

SURROUND-SOUND MIKING TECHNIQUES

So far we've examined two-channel stereo recording techniques. These methods reproduce the instruments and hall reverb in front of the listener, in the area between the two loudspeakers. In contrast, surround-sound places audio images around the listener. The musical ensemble is usually up front, and the hall ambience is all around.

Stereo uses two channels feeding two loudspeakers. Surround sound uses multiple channels feeding multiple speakers.

Surround gives a wonderfully spacious effect. It puts you inside the concert hall with the musicians. You and the music occupy the same space—you're part of the performance. For this reason, surround is more musically involving, more emotionally intense, than regular stereo.

A magazine devoted to multichannel sound production is *Surround Professional*, available on the Web at www.surroundpro.com/.

Here are the basic steps to create a surround recording:

1. Set up four or five mics in a surround array (described later in this chapter) and plug them into mic preamps or a mixer.
2. Record the amplified mic signals onto a multitrack hard-drive (HD) recorder.

3. Master a surround DVD from the four or five channels. You need DVD-Audio authoring software, a DVD-R recorder, and blank DVD discs. Two major titles of authoring software are DVD-Audio Creator by Sonic Solutions (www.sonic.com) and discWelder Bronze by Minnetonka software (www.minnetonka.com or www.discwelder.com).

4. Alternatively, master the surround recordings on super-audio compact discs or DVD video discs.

Surround Speaker Arrangement

In a technique inherited from the film industry, 5.1 surround-sound uses six channels feeding six speakers placed around the listener. This forms a 5.1 surround system, where the "point 1" is the subwoofer or low-frequency effects (LFE) channel. The LFE channel is band-limited to 120 Hz and below.

The six speakers are:

- Left front
- Center
- Right front
- Left surround
- Right surround
- Subwoofer

Figure 10-1 shows the recommended placement of monitor speakers for 5.1 surround sound. It is the standard setup proposed by the International Telecommunication Union (ITU). From the center speaker, the left and right speakers should be placed at ±30°, and the surrounds at ±110°.

The left- and right-front speakers provide regular stereo. The surrounds provide a sense of envelopment due to room ambience. They also allow sound images to appear behind the listener. Deep bass is filled in by the subwoofer. Because we do not localize low frequencies below about 120 Hz, the sub can be placed almost anywhere without degrading localization.

In a system originally developed for theaters, the center-channel speaker is mounted directly in front of the listener. In a home-theater system, it is placed just above the TV screen, or just below and in front of the TV screen. This speaker plays center-channel information in mono, such as dialog.

Why use a center speaker, when two stereo speakers create a phantom center image? If you use only two speakers and you sit off-center, the phantom image shifts toward the side on which you're sitting. But a

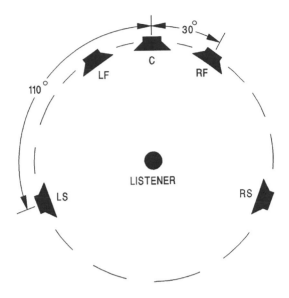

Figure 10-1 Recommended placement of monitor speakers for 5.1 surround sound.

center-channel speaker produces a real image, which does not shift as you move around the listening area.

Also, the phantom center image does not have a flat frequency response, but a center speaker does. Why is this? Remember that a center image results when you feed identical signals to both stereo speakers. The right-speaker signal reaches your right ear, but so does the left-speaker signal after a delay around your head. The same thing occurs symmetrically at your left ear. Each ear receives direct and delayed signals, which interfere and cause phase cancellations at 2 kHz and above. A center-channel speaker does not have this response anomaly.

With a phantom center image, the response is weak at 2 kHz because of the phase cancellation just mentioned. To compensate, recording engineers often choose mics with a presence peak in the upper midrange for vocal recording. The center-channel speaker does not need this compensation.

For sharpest imaging and continuity of the soundfield, all the speakers should be:

- the same distance from the listener;
- the same model (except the sub);
- the same polarity;

- direct-radiator types;
- driven with identical power amps;
- matched in sound pressure level with pink noise.

Typically the speakers are 4–8 feet from the listener and 4 feet high. Use a length of string to place the monitors the same distance from your head. The sub can go along the front wall on the floor.

Be sure that all the speakers sound the same so there is no change in tonal balance as you pan images around.

Surround-Sound Mic Techniques

You can record classical music in surround using four or five microphones, which capture the three-dimensional spatial character of the concert hall in which the musical ensemble is playing. Each mic feeds a separate track of a multitrack recorder.

Surround mic techniques are somewhat different from stereo mic techniques. In addition to the usual front-left and -right mics, you need two surround mics to pick up the hall ambience and a center mic to feed to the center channel. You don't need a subwoofer channel with this type of recording. Note that listening in surround reduces the stereo separation (stage width) because of the center speaker, but mic techniques for surround are optimized to counteract this effect.

A number of mic techniques have been developed for recording in surround. Let's take a look at them.

SoundField 5.1 Microphone System

This system is a single, multiple-capsule microphone (SoundField ST250 or MKV) and SoundField Surround Decoder for recording in surround. The decoder translates the mic's B-format signals (X, Y, Z, and W) into L, C, R, LR, RR, and mono subwoofer outputs. For details, see Chapter 12 under the heading "Surround Microphones" and Appendix B under the heading "SoundField Microphone."

Delos VR² Surround Miking Method

John Eargle, Delos's director of recording, developed Delos's VR^2 (Virtual Reality Recording) format.

In making these recordings, Eargle typically uses the mic placement shown in Figure 10-2. This method employs an Office de Radiodiffusion

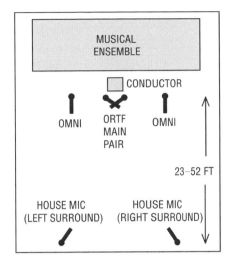

Figure 10-2 A Delos surround miking method.

Television Française (ORTF) pair in the center, flanked by two spaced omnis typically 12 feet apart. Two house mics (to pick up hall reverb) are placed about 23–52 feet behind the main pair. These house mics are cardioids aiming at the upper rear corners of the hall, spaced about 12 feet apart and about 30 feet high. Spot mics (accent mics) are placed within the orchestra to add definition to certain instruments.

The mics are assigned to various tracks of a digital multitrack recorder:

- 1 and 2: A mix of the ORTF-pair mics, flanking mics, house mics, and spot mics.
- 3 and 4: ORTF-pair mics.
- 5 and 6: Flanking mics.
- 7 and 8: House mics (surround mics).

NHK Method

The Japanese NHK Broadcast Center has worked out another surround miking method. Engineers found that, for surround recording, cardioid mics record a more natural amount of reverb than omni mics. The mics are placed as described below:

- A center-aiming mic feeds the center channel.
- A near-coincident pair feeds front left and front right.

145

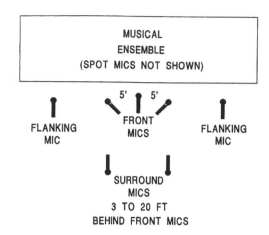

Figure 10-3 An NHK surround-sound miking method.

- Widely spaced flanking mics add expansiveness.
- Up to three pairs of ambience mics aim toward the rear.

Figure 10-3 shows the mic placement. The front mics are placed at the critical distance from the orchestra, where the direct-sound level matches the reverberant-sound level. Typically, this point is 12–15 feet from the front of the musical ensemble and 15 feet above the floor.

NHK engineers make this recommendation: when you're monitoring the surround program, the reverb volume in stereo listening should match the reverb volume in multichannel listening. That is, when you fold down or collapse the monitoring from 5.1 to stereo, the direct/reverb ratio should stay the same.

KFM 360 Surround Miking System

Jerry Bruck of Posthorn Recordings developed this elegant surround miking method. It is a form of the mid–side (MS) stereo technique.

Bruck starts with a modified Schoeps KFM 6U stereo microphone, which is a pair of omni mics mounted on opposite sides of a 7-inch hard sphere. Next to those mics, nearly touching, are two figure-eight mics, one on each side of the sphere, each aiming front and back (Figure 10-4). This array creates two MS mic arrays aimed sideways in opposite directions. The mics do not seriously degrade one another's frequency response.

In the left channel, the omni and figure-eight mic signals are summed to give a front-facing cardioid pattern. They are also differenced to give a

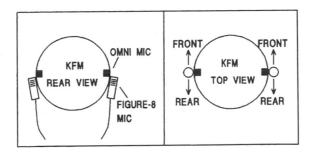

Figure 10-4 The KFM 360 surround miking system.

rear-facing cardioid pattern. The same thing happens symmetrically in the right channel. The sphere, acting as a boundary and a baffle, "steers" the cardioid patterns off to either side of center and makes their polar patterns irregular.

By adjusting the relative levels of the front and back signals, the user can control the front/back separation. As a result, the mic sounds like it is moving closer to or farther from the musical ensemble.

According to Bruck:

The system is revelatory in its ability to recreate a live event. Perhaps most remarkable is the freedom a listener has to move around and select a favored position, as one might move around in a concert hall to select a preferred seat. The image remains stable, without a discernible "sweet spot." The reproduction is unobtrusively natural and convincing in its sense of "being there."

The four mic signals can be recorded on a four-track recorder for later matrixing. The figure-eight mics need some equalization (EQ) to compensate for their low-frequency rolloff and loss in the extreme highs. To maintain a good signal-to-noise ratio, this EQ can be applied after the signal is digitized.

Five-Channel Microphone Array with Binaural Head

This method was developed by John Klepko of McGill University. It combines an array of three directional mics with a two-channel dummy head (Figure 10-5):

- *For the front-left and -right channels*: Identical supercardioid mics.
- *For the center channel*: A cardioid mic.

147

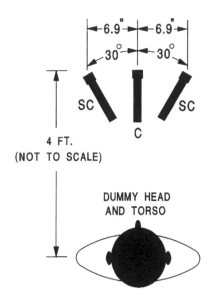

Figure 10-5 The Klepko surround-sound miking method.

- *For the surround channels*: A dummy head with two omni mics fitted into the ear molds.

The mics are shock mounted and have equal sensitivity and equal gains. Supercardioids are used for the front-left/right pair to reduce center-channel buildup. Although the dummy head's diffraction causes peaks and dips in the response, it can be equalized to compensate. During playback, the listener's head reduces the acoustical crosstalk that would normally occur between the surround speakers.

According to Klepko:

The walkaround tests form an image of a complete circle of points surrounding the listening position. Of particular interest is the imaging between ±30° and ±90°. The array produces continuous, clear images where other (surround) techniques fail.

The proposed approach is downward compatible to stereo, although there will be no surround effect. However, stereo headphone reproduction will resolve a full surround effect due to the included binaural head-related signals. Downsizing to matrix multichannel (5-2-4 in this case) is feasible, except that it will not properly reproduce binaural signals to the rear because of the mono surrounds. As well, some of the spatial detail recorded by the dummy-head microphone will be lost due to the usual bandpass

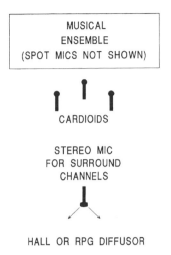

Figure 10-6 A DMP surround miking method.

filtering scheme (100 Hz–7 kHz) of the surround channel in such matrix systems.

DMP Method

DMP engineer Tom Jung has recorded in surround using a Decca Tree stereo array for the band and a rear-aiming stereo pair for the surround ambience (Figure 10-6). Spot mics in the band complete the miking. The Decca Tree uses three omni mics spaced a few feet apart, with the center mic placed slightly closer to the performers. It feeds the center channel in the 5.1 system.

The rear-aiming mics are either a coincident stereo mic, another Decca tree, or a spaced pair whose spacing matches that of the Decca tree outer pair. Jung tries to aim the rear mics at irregular surfaces to pick up diffuse sound reflections.

Woszczyk Technique

A recording instructor at McGill University, Wieslaw Woszczyk developed an effective method for recording in surround that also works well in stereo. The orchestra is picked up by a Pressure Zone Microphone (PZM) wedge made of two 18-inch × 29-inch hard baffle boards angled 45°. A mini omni mic is mounted on or flush with each board. At least 20 feet behind the wedge are the surround mics: two coincident cardioids angled 180° apart, aiming left and right, and in opposite polarity (Figure 10-7).

149

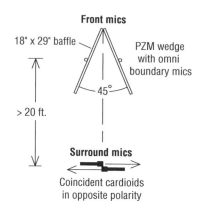

Figure 10-7 The Woszczyk surround miking method.

According to Woszczyk, his method has several advantages:

- Imaging is very sharp and accurate, and spaciousness is excellent due to strong pickup of lateral reflections.
- The out-of-phase impression of the surround pair disappears when a center coherent signal is added.
- The system is compatible in surround, stereo, and mono. In other words, the surround signals do not phase-interfere with the front-pair signals. That is because (1) the surround signals are delayed more than 20 milliseconds, (2) the two mic pairs operate in separate sound fields, and (3) the surround mics form a bidirectional pattern in mono, with its null aiming at the sound source.

If a PZM wedge is not acceptable because of its size and weight, other arrays with wide stereo separation may be substituted.

Williams Five Cardioid Mic Array

Michael Williams, an independent audio consultant, worked out the math to determine the best cardioid microphone arrangement for realistic reproduction of surround-sound fields. His method is shown in Figure 10-8.

Double MS Technique

Developed by Curt Wittig and Neil Muncy, the double MS technique uses a front-facing MS mic pair for direct sound pickup and a rear MS pair facing

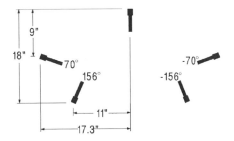

Figure 10-8 The Williams five cardioid mic array (MMA or Multichannel Microphone Array). See www.microphone-data.com.

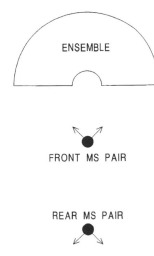

Figure 10-9 The double MS technique.

away from the front (Figure 10-9). The rear pair is placed at or just beyond the critical distance of the room, where the reverberant-sound level equals the direct-sound level. The matrixed outputs feed front-left, front-right, rear-left, and rear-right speakers. No center channel mic is specified, but you could use the front-facing cardioid mic of the front MS pair for this purpose.

Surround Ambience Microphone Array

Surround Ambience Microphone (SAM) array was developed by Gunther Theile of the Institute für Rundfunktechnik (IRT). Four cardioid mics are placed at 90° to each other and 21–25 cm apart. No center channel is described.

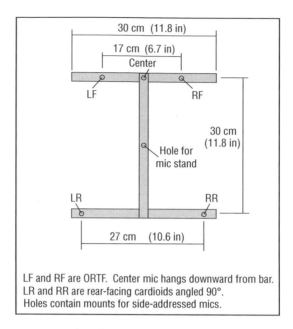

Figure 10-10 Mounting bar for Burmajster surround array.

Chris Burmajster Array

Based on extensive listening tests, this array was invented by Chris Burmajster of Innocent Ear Ltd. It includes an ORTF pair for left- and right-front channels, a center mic, and two rear-facing cardioids angled 90° for the left- and right-rear channels. The mics are mounted on the metal bar shown in Figure 10-10 (not commercially available). According to the inventor, this arrangement provides solid central imaging. The rear ambience channels sound coherent with the front channels, rather than "disjointed" as can happen with mics far back in the hall. Details are at http://homepage.ntlworld.com/chris.burmajster.

Ideal Cardioid Arrangement

This system uses a special mic mount with five arms that radiate out from a center point, like a star. At the end of each arm is a condenser mic aiming outward from the center. Two examples: The Microtech Gefell INA 5 uses five M930 mics in shock mounts (www.microtechgefell.de). In the SPL Atmos 5.1 Surround Recording System, five Brauner condenser mics feed a five-channel mixing console, which adjusts the mic polar patterns and

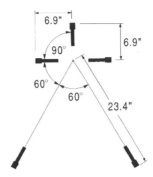

Figure 10-11 ICA used in the Brauner SPL Atmos 5.1 adjustable surround microphone and the Microtech Gefell INA 5 surround microphone.

offers panning, bass management, and surround monitoring. SPL's Website is www.spl-usa.com. Both systems use the Ideal Cardioid Arrangement (ICA 5, ITU-775 specification, Figure 10-11) developed by Volker Henkels and Ulf Herrmann.

Holophone H2-PRO Surround Mic

This is a surround microphone using several omni mic capsules flush-mounted in a football-shaped surface (see Figure 12-6). It captures up to eight channels of discrete surround sound and has eight XLR connectors (www.theholophone.com).

Sonic Studios DSM-4CS Four-Channel Surround Dummy Head

The Sonic Studios Website (www.sonicstudios.com) offers a four-mic array that you can put on your head, or on a dummy head, to record surround.

Slotte Method

Benedict Slotte developed a surround miking technique intended to produce very sharp images. The front three mics (Figure 10-12) are an optimized near-coincident array using three supercardioid mics. The array was designed for minimal crosstalk and time differences between mics. The level and time differences between the left-center pair are zero for a sound source at the midpoint between them. The same is true for the right-center pair. In his

153

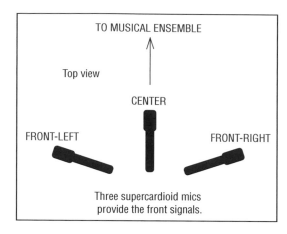

Figure 10-12 The Slotte method.

Audio Engineering Society (AES) paper (see Recommended Reading), Slotte lists a range of mic angling, spacing, and relative levels that work well.

Martin Method

Geoff Martin invented a surround technique using two Blumlein pairs: one for front left-right and the other for surround left-right (Figure 10-13). The goal is high interchannel coherence for direct sounds and low interchannel coherence for reverberation. Not shown is the center-channel mic, a figure-eight that is coincident with the front pair. The center mic can aim straight down to minimize image distortion, or it can aim forward but located at least 2.4 in (6 cm) below the main pair. Forward aiming is recommended if you want a lot of direct sound in the center speaker.

Mike Sokol's FLuRB Array

This array uses four coincident cardioid mics at 90° to each other aiming to the front, left, right, and back (Figure 10-14). The four mic signals feed a matrix processor that delivers the correct signals for 5.1 surround, up to 8.1 surround. The array is compact, relatively low cost, and convenient to use. Plus, it will sum to stereo or mono without phase cancellations. A matrix used with the FLuRB array is described at this Website: www.manleylabs.com/containerpages/flurb.html.

154

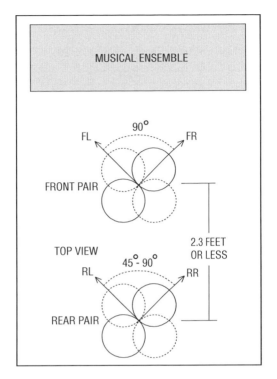

Figure 10-13 The Martin method.

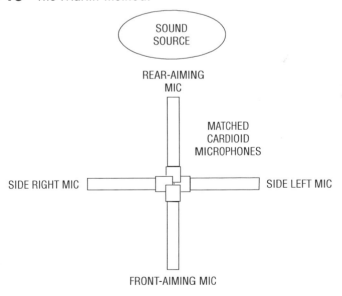

Figure 10-14 Mike Sokol's FLuRB Array.

Stereo Pair plus Surround Pair

In this method, the center-channel mic is omitted. You use a standard stereo pair of your choice to pick up the musical ensemble, plus another stereo pair of your choice to pick up the hall ambience. The hall mics feed the left- and right-surround channels. For example, the system might include two stereo mics placed back-to-back, separated by several feet.

You might try a hybrid approach for a pop-music concert: feed the front speakers a mix of multiple close-up mics on stage, and feed the rear speakers the signals from a rear-aiming stereo mic.

Recommended Reading

The Website www.dpamicrophones.com has a section called "Microphone University." In this section are several excellent papers on surround microphone techniques, many of which are not covered here.

Michael Williams and Guillaume Le Dû. "The Quick Reference Guide to Multi-channel Microphone Arrays, Part 1: Using Cardioid Microphones." Preprint 5336 from the 110th Convention of the Audio Engineering Society, May 2001.

Paul Segar and Francis Rumsey. "Optimisation and Subjective Assessment of Surround Sound Microphone Arrays." Preprint 5368 from the 110th Convention of the Audio Engineering Society, May 2001.

Benedict Slotte. "Sharpening the Image in 5.1 Surround Recording." Preprint 6509 from the 118th Convention of the Audio Engineering Society, May 2005.

Geoff Martin. "A New Microphone Technique for Five-Channel Recording." Preprint 6427 from the 118th Convention of the Audio Engineering Society, May 2005.

11

TROUBLESHOOTING
STEREO SOUND

Suppose that you're monitoring a stereo recording in progress or listening to a recording you've already made. Something doesn't sound right. How can you pinpoint what's wrong, and how can you fix it?

This section lists several procedures to solve audio-related problems. Read down the list of bad sound descriptions until you find one matching what you hear, then try the solutions until your problem disappears.

Before you start, check for faulty cables and connectors. Also check all control positions, rotate knobs, and flip switches to clean the contacts.

Distortion in the Microphone Signal

- Use pads or turn down the gain trims in your mixer.
- Switch in the pad in the condenser microphone, if any.
- If your recorder has a pad, switch it in. If your recorder has a mic gain switch, set it to the low-gain setting.
- Use a microphone with a higher "maximum SPL" specification.

Too Dead (Not Enough Reverberation)

- Place microphones farther from performers. *Play CD track 25 to hear how miking distance affects the amount of reverberation and depth.*
- Use omni-directional microphones.
- Record in a concert hall with better acoustics (longer reverberation time).
- Add artificial reverberation.
- Add plywood or plastic sheeting over the audience seats.
- If the venue acoustics are good, mix in a second pair of mics about 30 feet from the main pair.

Too Detailed, Close, or Edgy

- Place microphones farther from performers. *Play CD track 25 to hear how miking distance affects the amount of reverberation and depth.*
- Place microphones lower or on the floor (as with a boundary microphone).
- Using an equalizer in your mixing console, roll off the high frequencies.
- Use duller-sounding microphones.
- If using both a close-up pair and a distant ambience pair, turn up the ambience pair.
- If using spot mics, add artificial reverb or delay their signals to coincide with the main pair's signals.

Too Distant (Too Much Reverberation)

- Place microphones closer to performers. *Play CD track 25 to hear how miking distance affects the amount of reverberation and depth.*
- Use directional microphones (such as cardioids).
- Use a spaced pair of directional mics aiming straight ahead.
- Record in a concert hall that is less "live" (reverberant).
- If using both a close-up pair and a distant ambience pair, turn down the ambience pair.

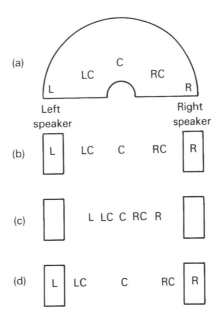

Figure 11-1 Stereo localization effects: (a) orchestra instrument locations (top view); (b) images accurately localized between speakers (the listener's perception); (c) narrow stage effect; and (d) exaggerated separation effect.

Narrow Stereo Spread

See Figure 11-1(c). *Play CD track 9 to hear narrow stereo spread*:

- Angle or space the main microphone pair farther apart.
- If doing mid–side stereo recording, turn up the side signal.
- Place the main microphone pair closer to the ensemble. *Note*: This also will make the ensemble sound closer.
- If monitoring with headphones, narrow stereo spread is normal when you use coincident techniques. Try monitoring with loudspeakers, or use near-coincident or spaced techniques.

Excessive Separation, Hole-in-the Middle, or Soloist Moves Too Much

See Figure 11-1(d). *Play CD track 20 to hear excessive separation*:

- Angle or space the main microphone pair closer together.

159

- If doing mid–side stereo recording, turn down the side signal or use a cardioid mid instead of an omni mid.
- In spaced-pair recording, add a microphone midway between the outer pair and pan its signal to the center.
- Place the microphones farther from the performers. *Note:* This also will make the ensemble sound more distant.
- Place the loudspeaker pair closer together. Ideally, they should be as far apart as you are sitting from them, to form a listening angle of 60°.

Poorly Focused Images

- Avoid spaced-microphone techniques. *Play CD tracks 6, 18–20, 22, and 24 to hear the image focus of some spaced-microphone techniques.*
- Use a microphone pair that is better matched in frequency response.
- If the sound source is out of the in-phase region of microphone pickup, move the source or the microphone. For example, the in-phase region of a Blumlein pair of crossed figure-eights is ±45° relative to center.
- Be sure that each spot mic is panned so that its image location coincides with that of the main pair.

Maybe the image-focus problem is in your monitoring system instead of the mic technique. Try these suggestions:

- Use loudspeakers designed for sharp imaging. Usually these are signal-aligned, have vertically aligned drivers, have curved edges to reduce diffraction, and are sold in matched pairs.
- Place the loudspeakers several feet from the wall behind them and from side walls to delay and weaken the early reflections that can degrade stereo imaging.
- Put acoustic foam on the wall behind the speakers to absorb early reflections.
- Use Nearfield monitors (about 3 feet apart and 3 feet from you).

Images Shifted to One Side (Left-Right Balance Is Faulty)

- Adjust the right-or-left recording level so that center images are centered.

- Use a microphone pair that is better matched in sensitivity.
- Aim the center of the mic array exactly at the center of the ensemble.
- Sit exactly between your stereo speakers, equidistant from them. Adjust the balance control or level controls on your monitor amplifier to center a mono signal. *Play CD tracks 1–4 to set up your speakers correctly for stereo listening.*

Lacks Depth

Play CD track 25 to hear a voice reproduced with gradually increasing depth:

- Avoid spot mics.
- If you must use spot mics, keep their level low in the mix, and delay their signals to coincide with those of the main pair.
- The problem might be your monitor speakers. Use monitors with signal-aligned drivers, put acoustical absorption on the wall behind the speakers, and space the speakers several feet from the wall behind them.

Lacks Spaciousness

- Use spatial equalization (described in Appendix A under the heading "Spaciousness and Spatial Equalization").
- Space the microphones apart.
- Place the microphones farther from the ensemble. *Play CD track 25 to hear how miking distance affects the sense of spaciousness.*
- Mix in a distant mic pair placed about 30 feet back in the hall.
- Use a spaced pair or near-coincident pair instead of a coincident pair.
- Record in a venue with stronger side reflections or longer reverberation time.
- Add artificial reverb.
- See the suggestions under the heading "Early Reflections Too Loud" below.

Early Reflections Too Loud

- Place mics closer to the ensemble and add a distant microphone for reverberation (or use artificial reverberation).

161

- Place the musical ensemble in an area with weaker early reflections.
- If the early reflections come from the side, try aiming bidirectional mics at the ensemble. Their nulls will reduce pickup of side-wall reflections.

Bad Balance (Some Instruments Too Loud or Too Soft)

- Place the microphones higher or farther from the performers.
- Move quiet instruments closer to the stereo pair and vice versa.
- Ask the conductor or performers to change the instruments' written dynamics.
- Add spot microphones close to instruments or sections needing reinforcement. Mix them in subtly with the main microphones' signals.
- Increase the angle between mics to reduce the volume of center instruments and vice versa.
- If using a mid–side mic and matrix, vary the mid–side ratio to slightly change the balance of middle instruments to side instruments. This will also change the stereo spread.

Muddy Bass

- Aim the bass-drum head at the microphones.
- Put the microphone stands and bass-drum stand on resilient isolation mounts, or place the microphones in shock-mount stand adapters.
- Suggest that the tympani player use hard sticks instead of soft, and maybe play quieter.
- Roll off the low frequencies or use a high-pass filter set around 40–80 Hz.
- Record in a concert hall with less low-frequency reverberation.

Rumble from Air Conditioning, Trucks, and So On

- Temporarily shut off air conditioning. Record in a quieter location or at a quieter time of day.

- Use a high-pass filter set around 40–80 Hz. Use microphones with limited low-frequency response.
- Mike closer and add artificial reverberation.

Bad Tonal Balance (Too Dull, Too Bright, Colored)

- Try different microphones.
- If a microphone must be placed near a hard reflective surface, use a boundary microphone to prevent phase cancellations between direct and reflected sounds.
- Adjust equalization. Compared to omni condenser mics, directional mics usually have a rolled-off low-frequency response and may need some bass boost.
- If strings sound strident, move mics farther away or lower.
- If the tone quality is colored only in mono monitoring, use coincident-pair techniques.

12

STEREO, SURROUND, AND BINAURAL MICROPHONES AND ACCESSORIES

This is a listing of stereo microphones, surround microphones, headworn mics, dummy heads, mid–side (MS) matrix boxes, and stereo microphone stand adapters. The list is up to date only for the date of publication of this book. Since models and prices change, please contact the manufacturers for current information. This chapter is not a catalog of preferred products, but rather an illustration of currently available products and their features. Note the following definitions of microphone specifications:

- *Side-addressed*: The axis of maximum sensitivity is at right angles to the microphone's long axis. You aim the side of the mic at the sound source.
- *End-addressed*: The axis of maximum sensitivity is the same as the mic's long axis. You aim the front of the mic at the sound source.
- *Binaural*: Two small mics are mounted flush with each ear canal. They pick up the "sound shadowing" effect of the head on the spectrum of the sound source (the frequency response and phase response of the head), plus the acoustic effects of the pinnae or outer ears.

- *HRTF (head-related transfer function)*: Two small mics are mounted on the temples. They pick up the effect of the head on the spectrum of the sound source (the frequency response and phase response of the head), but not the effects of the pinnae.

Stereo Microphones

AKG C-426B Comb: Two twin-diaphragm condenser capsules, one atop the other, for MS or XY use. Three polar patterns with six intermediate steps. Low-cut switch, attenuator, and shock mount. Side-addressed. See Figure 12-1.

Figure 12-1   AKG C-426B Comb stereo microphone (courtesy: AKG Acoustics, Inc.).

Audio-Technica AT822: End-addressed XY mic. Compact and lightweight. Switchable low-frequency roll off, battery powered only, unbalanced outputs. Several accessories.

Audio-Technica AT825: Same as above but with battery or phantom power, balanced outputs, and flatter response. See Figure 12-2.

Audio-Technica ATR25: End-addressed XY mic designed for video cameras. On–off switch. Several mounts and cables.

Audio-Technica Pro 24: Low-cost XY mic works on DC bias from recorder. End-addressed.

Beyerdynamic MC 833: Three large-diaphragm capsules with variable positions. Works for MS or XY recording without an external matrix. For orchestra/recital recording, ambience, and sampling. End-addressed.

Church Audio B-99M mini stereo mic and B-99A larger stereo mic: Stealth headworn cardioid mics. All plug into a Church Audio mic preamp or stereo recorder with a stereo mini-phone input.

Core Sound Jecklin Disk (Optimal Stereo Signal (OSS) system): Padded baffle mount for two ear-spaced omni mics (Figure 12-3).

Core Sound Schneider Disk: Has foam hemispheres covering the baffle. Compared to the Jecklin Disk, offers slightly better approximation of binaural recording when played over headphones.

Figure 12-2 Audio-Technica AT825 end-addressed XY mic (courtesy: Audio-Technica).

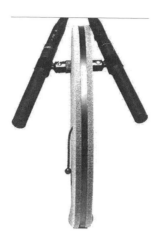

Figure 12-3 Jecklin Disk (courtesy: Josephson Engineering).

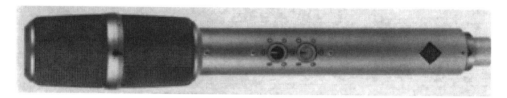

Figure 12-4 Neumann USM69i stereo microphone (courtesy: Neumann USA).

Crown SASS-P MKII Pressure Zone Microphone (PZM) Stereo Microphone: Two ear-spaced PZMs on angled boundaries. For near-coincident, mono-compatible recording. End-addressed. See Figure C-3.

Josephson OSSD 1SK Jecklin Disk: Stereo mic baffle.

Josephson SPB: Spherical baffle for 21 mm (0.8 in) microphones.

Neumann USM69i: Multi-pattern dual stereo MS/XY microphone. Upper capsule can be rotated relative to the lower one through 270°. Side-addressed. See Figure 12-4.

Neumann KFM 100: A 20 cm (7.8-inch) wooden sphere with two diametrically opposed pressure mics flush-mounted in the sphere. Time and spectral differences between channels produce the stereo imaging, with excellent depth reproduction. Response down to 10 Hz. Side-addressed.

Pearl DS60: Two rectangular dual-membrane capsules mounted one above the other, 90° apart. For XY or MS methods. At your mixing console you can select the desired polar pattern and stereo system: cardioid,

Figure 12-5 Rode NT4 XY stereo microphone (courtesy: Rode Microphones).

figure-eight, omni, XY, MS, or Blumlein. Phantom powered. Four XLR connectors. Side-addressed.

Pearl TL4/TL44: For mono or stereo use. Two discrete cardioid outputs may be used together or apart to get cardioid, omni, figure-eight, or 180° XY by adjusting your mixing-console controls. Phantom powered. Side-addressed.

Rode NT4: XY stereo microphone, battery or phantom powered (Figure 12-5).

Royer SF-12: Stereo coincident ribbon microphone for Blumlein or MS technique.

Royer SF-24: Phantom-powered stereo ribbon microphone for Blumlein or MS technique.

Sanken CMS-7S: Portable MS mic. Requires external matrix box. Fixed angle of polar patterns. CMS-7 has cardioid mid unit; CMS-7H has hypercardioid mid unit. End-addressed.

Sanken CMS-9: Portable MS mic similar to CMS-7, but with internal matrix so no external box is required. Left/right or MS outputs, fixed angle of polar patterns. End-addressed.

Sanken CUW-180: Adjustable XY microphone. End-addressed.

Schoeps KFM6 Sphere Microphone: Two omni capsules flush-mounted in an 8-inch wooden globe. Head-related phase and frequency compensation. Phantom powered. Side-addressed.

Schoeps KFM 360 Sphere Microphone: Smaller version of Schoeps KFM6 Sphere Microphone for surround use (see Figure 10-4).

Schoeps MSTC 64G Office de Radiodiffusion Television Française or "ORTF" Microphone: Two MK 4 cardioid capsules at either end of a T-shaped dual-amplifier body, spaced 17 cm (6.7 inches) apart and angled 110°. XLR-5M output, 12–48 V phantom powered.

Schoeps CMXY 4Vg: Two XY side-addressed cardioid CCM 4Vg capsules, interlocked so that they rotate in opposite directions to adjust stereo spread from 0° to 180°. Compact, with swivel base for hanging, boom mounting, or tabletop use. 12–48 V phantom powered. Choice of connectors.

Two CMBI/MK or CMC/MK mics, with Schoeps accessories, can be set up for compact XY, MS, or near-coincident pickups.

Schoeps XY system: UMS 20 stereo bar with MK 4 capsules and CMC 6 microphone amplifier.

Schoeps Blumlein system: Two MK 8 capsules with stereo microphone bar.

Schoeps MS system: AMS 22 elastic suspension for MS. Stereo microphone bar UMS 20 with MK 8 and MK 4 or MK 41 microphone capsules and CMC 6 microphone amplifier.

Schoeps ORTF system: UMS 20 stereo bar with MK 4 capsules and CMC 6 microphone amplifier. Or STC stereo bar for ORTF with CCM 4 compact microphone or MK 4 microphone capsule plus KC cable and CMC microphone amplifier.

Shure VP88: MS mic with stereo or MS outputs. Switchable stereo spread control on microphone. Low-cut switch. End-addressed.

Sony ECM-DS70P: Mini stereo mic with two cardioid capsules back-to-back. 1/8-inch (.317 cm) stereo phone plug built-in. 1-m (3.3-ft) detachable mic cable. Plug-in power operation. Response 100 Hz–15 kHz.

Sony ECM-MS907: MS stereo mic with 1/8-inch stereo phone plug on 2-m (6.6-ft) cable. Battery powered. Adjustable stereo spread. Response 100 Hz–15 kHz.

Sony ECM-MS957: MS stereo mic with 1/8-inch (.317 cm) stereo phone plug on 2-m (6.6-ft) cable. Battery powered. Adjustable stereo spread. Response 50 Hz–18 kHz.

SoundField Microphones: See SoundField under the heading "Surround Microphones" in this chapter.

The Sound Professionals offer a wide variety of stereo mics at different price/quality levels. For example, the SP-SPSM-6 is a premium handheld/stand stereo mic with cardioid or omni mic capsules on short boom arms. The SP-SPSM-3 is a small, T-shaped plug-in stereo mic. The SP-PSM3 is an MS stereo mic like the Sony ECM-MS907.

Studio Projects LSD2: Large-diaphragm stereo condenser mic. Rotatable capsules, switchable polar patterns, XY or MS, side-addressed.

T.H.E. Audio BS-3D sphere microphone for stereo and binaural recording. Two flush-mounted condenser mics in a head-size wooden sphere.

Many handheld flash-memory recorders come with built-in or plug-in stereo microphones.

Surround Microphones

Holophone H2 Pro: Surround microphone captures up to 7.1 channels using an ellipsoid baffle with mic capsules. Discrete outputs; no matrix needed. See Figure 12-6.

Holophone H3-D: Like the Holophone H2 Pro, but a less-expensive model with 5.1 channels.

Manley FLuRB processor for Mike Sokol's FLuRB Array: Sokol's mic technique uses four matched cardioid mics aiming front, left, right, and back in a coincident array. You plug the four mics into the Manley FLuRB matrix to get 5.1, 6.1, and 7.1 surround. The processor was in development when this book was published.

Microtech Gefell INA 5: This array uses five shock-mounted M930 mics in the Ideal Cardioid Arrangement (ICA 5, ITU-775 specification).

Schoeps Surround System: Includes KFM 360 Sphere Microphone, two CCM 8L figure-eight mics, and DSP-4 KFM 360 processor.

SoundField Microphone (ST250 or MKV): Uses four capsules arranged in a tetrahedron, phase-matrixed for true coincidence. Processor offers remote control of polar pattern, azimuth (horizontal rotation), elevation (vertical tilt), and dominance (fore/aft movement). AC or battery powered. Provides stereo left/right, MS, mono or 4-channel B-format signals that

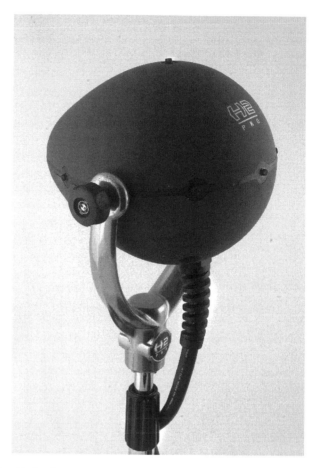

Figure 12-6 Holophone H2 Pro surround microphone (courtesy: Holophone).

can be matrixed into surround signals. Side-addressed. With flight case and 20-m (65.6 ft) cable.

SoundField 5.1 Microphone System: A single, multiple-capsule microphone (SoundField ST250 or MKV) and SoundField Surround Decoder for recording in surround. The decoder translates the mic's B-format signals (X, Y, Z, and W) into L, C, R, LR, RR, and mono subwoofer outputs.

SoundField MKV Microphone System: Multi-capsule microphone and calibrated 2U processor provides mono, stereo, M/S, and surround. Controls Azimuth (Rotate), Elevation (Tilt), and Dominance (Zoom) and Solo each mic capsule. B-format outputs and inputs.

172

SoundField SPS422: Less-expensive SoundField model. Two forward-facing capsules (left-front and right-front) and two backward-facing capsules (left-back and right-back). Processor offers remote control of pattern and width.

SoundField SPS422B Microphone System: Multi-capsule microphone and calibrated 1U processor provides mono, stereo, M/S, and surround. Adjustable mic parameters include Gain, End Fire, Invert, High-Pass, Polar Patterns, Stereo Width, and Headphone monitoring.

SoundField SP451 Surround Sound Processor: Rackmount device sends surround sound directly to digital recorders "live" as the performance takes place. In post, it outputs a 5.1 surround mix and stereo mix from a pre-recorded B-format program. Creates up to three surround mic arrays with various polar patterns. Has B-format inputs (W, X, Y, and Z) and eight surround outputs.

SoundField ST350 Portable Microphone System: Multi-capsule microphone and mic preamp/controller provides surround and stereo line-level signals. AC or battery powered.

SoundField Surround Zone: Digital audio workstation (DAW) plug-in has the features of both the SP451 Surround Processor and MKV System. Accepts B-format signals from any of the SoundField microphone models.

SPL Atmos 5.1 Surround Miking System, Model 9843: A system composed of the Atmos 5.1 Controller and the Brauner ASM 5 Adjustable Surround Microphone. The controller includes high-quality mic preamps with a master gain control, a surround panning matrix, subchannel and low-frequency effect (LFE), stereo in/out, polar-pattern control, stereo spread control, and 6-channel monitoring. The microphone array has five matched Brauner VM 1 mics mounted 17.5 cm (6.9 in) from the center in the ICA (ICA 5, ITU-775 specification).

Dummy Heads and Headworn Binaural Mics

Head Acoustics model HMS III: Dummy head with omni measurement microphones. A record processor equalizes the head signals to have flat response in a frontal free field. A reproduce unit contains free-field equalizers for use with Stax SR-Lambda Professional headphones. Also available for recording is a unit with Schoeps capsules, two XLR-type outputs, and switchable equalization (EQ) for a frontal free field or random-incidence SoundField.

AM3D Valdemar dummy head. Two flush-mounted mics in pinnae. Includes torso.

Figure 12-7 Core Sound CSB1 binaural microphones (courtesy: Core Sound).

Bruel & Kjaer 4100 Head and Torso Simulator: Uses two B&K mic capsules in a detailed replica of a human head and torso.

Church Audio binaural mics: Mics powered by a battery box or DC bias from a recorder or mic preamp. Church Audio's 1/2-inch mics cost less than their smaller "stealth" mics. They also offer an ST-20A battery-powered mic preamp with a stereo mini-phone input and output.

Core Sound binaural mics: A variety of binaural microphone sets that you wear on your head (Figure 12-7). Omni or cardioid capsules. Powered by a battery box or DC bias (plug-in power) from a recorder. High-end models use DPA 4060/4061 capsules. Stealthy Cardioid set uses cardioid capsules to reduce pickup of room acoustics.

Sennheiser binaural mics sold by Microphone Madness: A number of binaural microphones such as the MM-HLSC-2 cardioid model and the MM-BSM-9 omni model.

Neumann KU 100 "Fritz III" dummy head binaural system: A detailed human-head replica with omni microphones inside the ears. Loudspeaker compatible. For music recording, radio drama, film special effects, outdoor nature recordings, acoustic evaluation, and scientific research. Powered by internal batteries or external phantom-power supply. See Figure 12-8.

Sonic Studios Dimensional Stereo Mics (DSM): Unlike binaural mics, which are worn in the ears, DSM mics are two mini omni condenser mics meant to be worn on your temples (just forward of the ears). They are based on HRTF operation rather than binaural operation. DSM mics are said to provide more realistic sound over loudspeakers than binaural mics. Products include: DSM headworn mics, DSM HRTF baffle-mounted mics, DSM-15/x and 66/x stereo headworn mic, and 4-channel surround DSM mic system. Although they are normally headworn, DSM mics can

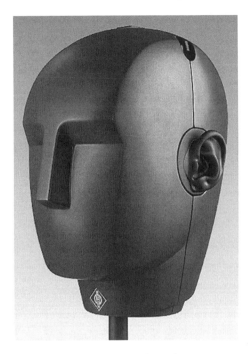

Figure 12-8 Neumann KU 100 dummy head (courtesy: Neumann USA).

be mounted on the Sonic Studios GUY or LiteGUY HRTF baffles, which are something like a dummy head but made of absorbent Sorbothane.

Soundman OKM binaural mics and OKM dummy head.

T.H.E. Audio BS-3D Sphere Microphone for stereo and binaural recording. Two flush-mounted condenser mics in a head-size wooden sphere.

Stereo and Surround Microphone Adapters

Audio Engineering Associates (AEA) Stereo Microphone Positioner (SMP) (Figure 12-9): Positions coincident and near-coincident arrays. Handles large mics, such as Coles 4038. Vertical or horizontal arrays, stand mounted or hung. Rotation angles are lines engraved at ±30°, 45°, and 55°. Center-to-center spacing is marked on a ruler with ORTF position shown. Various lengths available.

AEA SMT4038: Stereo template and bar.

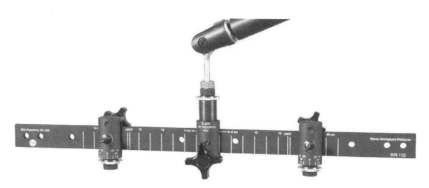

Figure 12-9 AEA SMP (courtesy: AEA).

AEA Tree Series: Mini-Tree, Decca Tree, and Super Tree. These multiple-microphone array mounts are used to securely position heavy microphones. The AEA modular microphone array can be configured to hold microphones for 5.1-channel recording.

Audio-Technica AT 8450: Stereo Hanger for two ES933 mics.

Beyer MAV 802: XY/ORTF mount.

Bruel & Kjaer CXY4000: Adjustable linear stereo rail for XY, MS, near-coincident, or spaced-pair miking.

Danish Pro Audio UAO 836: Stereo bar with sliding mic holders for A–B stereo. DUA 0019 19 mm (0.75 in) spacer allows XY and ORTF techniques.

Neumann DS 21 mt dual microphone mount: Can be used to combine two miniature microphones and two bent capsule-extension tubes into one fixed assembly for stereo recordings.

Neumann DS 120: Double mount with sliding mic-holder studs for XY/near-coincident miking.

Neumann DA-AK and DA-KM: Elastic suspension mounts for MS miking.

On-Stage MY-500 stereo microphone mount: Low-cost, non-sliding mic mounts for XY or ORTF.

Sabra Som ST2 stereo microphone mount: Sliding mic mounts for XY or ORTF.

Schoeps UMS 20: Universal stereo bracket with balljoint. Has two sliding, locking clamps on a crossbar, with detents set at the correct angles and positions for XY, MS, and ORTF. See Figure 12-10.

Schoeps MAB 1000: AB and ORTF mounting bar.

Schoeps M100C: Miniature ORTF mounting bar for MK4 capsules.

Figure 12-10 Schoeps UMS 20 stereo microphone adapter.

Schoeps STCg: ORTF T-bar with SG20, sets two cardioid capsules 17 cm (6.7 inches) apart and angled 110°.

Schoeps MS-BLM: Boundary-layer twin clamp for MS, mounts KC5 and MK8 on a BLM-3 plate.

Schoeps SGMSC: Mounting stud for MS setup of MK8 and MK−. With swivel, stand clamp, 3/8-inch adapter.

Schoeps AMSCI: Elastic suspension with swivel. Holds an MK cardioid and an MK figure-eight mic capsule parallel for MS pickup.

Shure A27M: Uses two rotating stacked cylinders. Lets you adjust angle and spacing for coincident and near-coincident methods. See Figure 12-11.

Sound Professionals SP-DTS-12: Mic mount for XY and ORTF. Contains non-sliding studs for mic holders.

MS Matrix Decoders

Groove Tubes MS Stereo Matrix Line Encoder: Encodes mid and side line-level signals into left and right signals. Balanced or unbalanced.

Schoeps VMS 5U: Stereo mic preamp with phantom power and MS matrix. Low-cut filter, balanced I/O, battery or external DC power.

SoundField: The SoundField Microphone systems described under "Surround Microphones" include MS decoders.

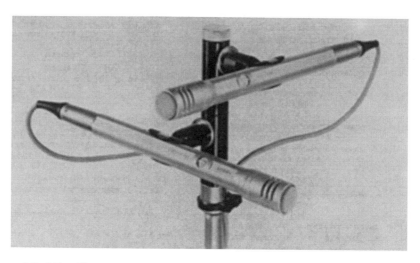

Figure 12-11 Shure A27M stereo microphone adapter (courtesy: Shure, Inc.).

Company Websites

AKG Acoustics US
www.akgusa.com
AM3D
www.am3d.com
Audio Engineering Associates (AEA)
www.wesdooley.com
Audio Technica US Inc.
www.audio-technica.com
Beyerdynamic Inc.
www.beyerdynamic.com
Bruel & Kjaer Instruments, Inc.
www.bkhome.com
Church Audio
www.church-audio.ca or
http://stereo.ebay.com/church-audio
Core Sound
www.core-sound.com
Crown International Inc.
www.crownaudio.com
Danish Pro Audio
dpamicrophones.com

Groove Tubes
www.groovetubes.com
HEAD Acoustics
www.head-acoustics.de
Holophone
www.holophone.com
Josephson Engineering
www.josephson.com
Manley Labs (FLuRB matrix)
www.manleylabs.com/containerpages/flurb.html
Microphone Madness
www.microphonemadness.com
Microtech Gefell
www.microtechgefell.de
Neumann USA
www.neumannusa.com
On-Stage MY500
(various vendors, such as www.bhphotovideo.com)
Pearl Microphones AB
www.pearl.se/
Rode Microphones
www.rode.com.au
Royer Microphone Labs
www.royerlabs.com
Sabra Som
www.sabra-som.com
Sanken Microphones
www.turneraudio.com
Schoeps Microphones
www.schoeps.de
Sennheiser Electronic Corp.
www.sennheiser.com
Shure Inc.
www.shure.com
Sonic Studios (DSM mics)
www.sonicstudios.com
Sony Professional Products
www.sony.com
SoundField Ltd.
www.soundfield.com

Soundman
www.soundman.de/englisch/english.htm
The Sound Professionals
www.soundprofessionals.com
SPL (Atmos 5.1 system)
www.spl-usa.com/Atmos/in_detail.html
Studio Projects
www.studioprojects.com
T.H.E. Audio
www.theaudio.com

A

STEREO IMAGING THEORY

These appendices are more academic than the rest of the book. They are for readers who want a deeper understanding of stereo mic techniques. You can use those techniques without reading this material. However, if you want to know how stereo works or to develop your own stereo array, it's worthwhile to study the theory and math in this appendix.

A sound system with good stereo imaging can form apparent sources of sound, such as reproduced musical instruments, in well-defined locations, and usually between a pair of loudspeakers placed in front of the listener. These apparent sound sources are called *images*.

This appendix explains terms related to stereo imaging, how we localize real sound sources, how we localize images, and how microphone placement controls image location.

Definitions

First, we'll define several terms related to stereo imaging. *Fusion* refers to the synthesis of a single apparent source of sound (an image or "phantom image") from two or more real sound sources (such as loudspeakers).

The *location* of an image is its angular position relative to a point straight ahead of a listener or its position relative to the loudspeakers. This is shown in Figure A-1. A goal of high fidelity is to reproduce the images in the locations intended by the recording engineer or producer.

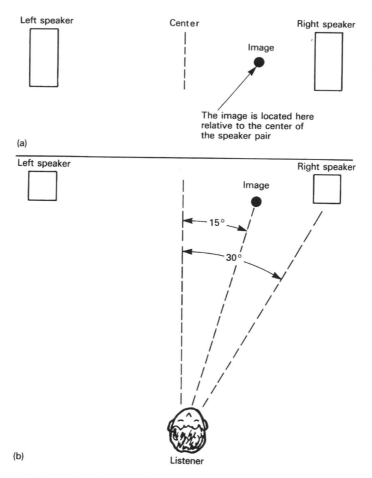

Figure A-1 Example of image location: (a) listener's view and (b) top view.

In some productions, usually classical music recordings, the goal is to place the images in the same relative locations as the instruments were during the live performance.

Stereo spread or *stage width* (Figure A-2) is the distance between the extreme-left and extreme-right images of a reproduced ensemble of instruments. The stereo spread is wide if the ensemble appears to spread all the way between a pair of loudspeakers. The spread is narrow if the ensemble occupies only a small space between the speakers. Sometimes the reproduced reverberation or ambience spreads from speaker to speaker

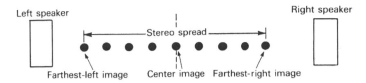

Figure A-2 Stereo spread or stage width.

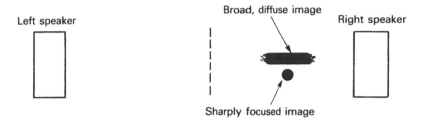

Figure A-3 Image focus or size (listener's perception).

even when the reproduced ensemble width is narrow. *Play CD tracks 14 and 17 to hear accurate stereo imaging with a wide spread. Play CD tracks 9 and 10 to hear a narrow stereo spread.*

Image *focus* or *size* (Figure A-3) refers to the degree of fusion of an image, its positional definition. A sharply focused image is described as being pinpointed, precise, narrow, sharp, resolved, well defined, or easy to localize. A poorly focused image is hard to localize, spread, broad, smeared, vague, and diffuse. A *natural image* is focused to the same degree as the real instrument being reproduced. *Play CD track 17 to hear sharply focused images and play CD track 22 to hear poorly focused images.*

Depth is the apparent distance of an image from a listener, the sense of closeness and distance of various instruments. *Play CD track 25 to hear how miking distance affects the sense of depth.*

Elevation refers to an image position above the line between the speakers.

Image movement is a reproduction of the movement of the sound source, if any. The image should not move unaccountably.

Localization is the ability of a listener to tell the direction of a sound. It is also the relation between interchannel or interaural differences and perceived image location. ("Interaural differences" means "differences between signals at the two ears.")

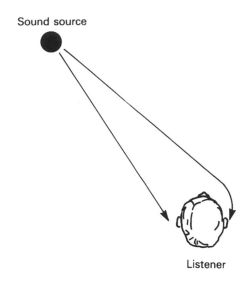

Figure A-4 Sound traveling from a source to a listener's ears.

How We Localize Real Sound Sources

The human hearing system uses the direct sound and early reflections to localize a sound source. The direct sound and reflections within about 2 milliseconds contribute to localization (Wallach et al., 1973; Bartlett, 1979). Reflections occurring up to 5–35 milliseconds after the direct sound influence image broadening (Gardner, 1973). Distance or depth cues are conveyed by early reflections (less than 33 milliseconds after the direct sound). Echoes delayed more than about 5–50 milliseconds (depending on program material) do not fuse in time with the early sound but contribute to the perceived tonal balance (Carterette and Friedman, 1978, pp. 62, 210).

Imagine a sound source and a listener. Let's say that the source is in front of the listener and to the left of center (as in Figure A-4). Sound travels a longer distance to the right ear than to the left ear, so the sound arrives at the right ear after it arrives at the left ear. In other words, the right-ear signal is delayed relative to the left-ear signal. Every source location produces a unique arrival-time difference between ears (Vanderlyn, 1979).

In addition, the head acts as an obstacle to sounds above about 1000 Hz. High frequencies are shadowed by the head, so a different spectrum (amplitude versus frequency) appears at each ear (Shaw, 1974; Mehrgardt and Mellert, 1977). Every source location produces a unique spectral difference between ears (Figure A-5).

184

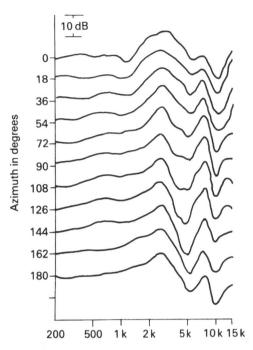

Figure A-5 Frequency response of the ear at different azimuth angles: 0° is straight ahead; 90° is to the side of the ear being measured; and 180° is behind the head (after Mehrgardt and Mellert).

We have learned to associate certain interaural differences with specific directions of the sound source. When presented with a new source location, we match what we hear with a memorized pattern of a similar situation to determine direction (Rumsey, 1989, p. 6).

As stated before, an important localization cue is the interaural arrival-time difference of the signal envelope. We perceive this difference at any change in the sound: a transient, a pause, or a change in timbre. For this reason, we localize transients more easily than continuous sounds (Rumsey, 1989, p. 3). The time difference between ear signals can also be considered as a phase difference between sound waves arriving at the ears (Figure A-6). This phase shift rises with frequency.

When sound waves from a real source strike a listener's head, a different spectrum of amplitude and phase appears at each ear. These interaural differences are translated by the brain into a perceived direction of the sound source. Every direction is associated with a different set of interaural differences.

185

The ears make use of interaural phase differences to localize sounds between about 100 and 700 Hz. Frequencies below about 100 Hz are not localized (making "subwoofer/satellite" speaker systems feasible; Harvey and Schroeder, 1961). Above about 1500 Hz, amplitude differences between ears contribute to localization. Between about 700 and 1500 Hz, both phase and amplitude differences are used to tell the direction of a sound (Eargle, 1976, Chapters 2 and 3; Cooper and Bauck, 1980).

Small movements of the head change the arrival-time difference at the ears. The brain uses this information as another cue for source location (Rumsey, 1989, p. 4), especially for distance and front/back discrimination.

The outer ears (pinnae) play a part as well (Gardner and Gardner, 1973). In each pinna, sound reflections from various ridges combine with the direct sound, causing phase cancellations and frequency notches in the perceived spectrum. The positions of the notches in the spectrum vary with the source height. We perceive these notch patterns not as a tonal coloration but as height information. Also, we can discriminate sounds in front from those in back because of the pinnae's shadowing effect at high frequencies.

Some of the cues used by the ears can be omitted without destroying localization accuracy if other cues are still present.

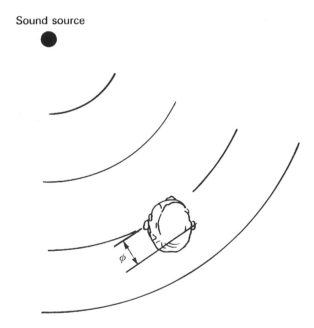

Figure A-6 Phase shift ϕ between sound waves at the ears.

How We Localize Images Between Speakers

Now that we've discussed how we localize real sound sources, let's look at how we localize their images reproduced over loudspeakers. Imagine that you're sitting between two stereo speakers, as in Figure A-7. If you feed a musical signal equally to both channels in the same polarity, you'll perceive an image between the two speakers. Normally, you'll hear a single synthetic source, rather than two separate loudspeaker sources. *Play CD track 1 to hear a phantom image in the center between your speaker pair.*

Each ear hears both speakers. For example, the left ear hears the left-speaker signal, then, after a short delay due to the longer travel path, hears the right-speaker signal. At each ear, the signals from both speakers sum or add together vectorially to produce a resultant signal of a certain phase.

Suppose that we make the signal louder in one speaker. That is, we create a level difference between the speakers. Surprisingly, this causes a signal time difference at the ears (Rumsey, 1989, p. 8). This is a result of

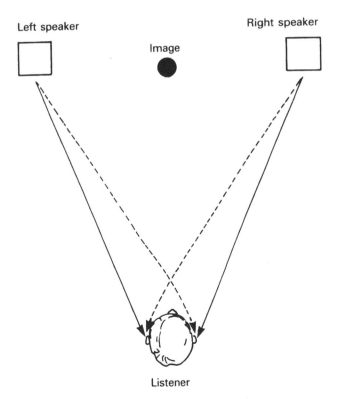

Figure A-7 Two ears receiving signals from two speakers.

the phasor addition of both speaker signals at each ear. For example, if the left-speaker signal is louder, the phase delay of the vector sum is higher in the right ear than in the left ear. So the right-ear signal is delayed with respect to the left-ear signal.

Remember to distinguish interchannel differences (between speaker channels) from interaural differences (between ears). An interchannel level difference does not appear as an interaural level difference, but rather as an interaural time difference.

We can use this speaker-generated interaural time difference to place images. Here's how: suppose we want to place an image 15° to one side. A real sound source 15° to one side produces an interaural time difference of 0.13 millisecond. If we can make the speakers produce an interaural time difference of 0.13 millisecond, we'll hear the image 15° to one side. We can fool the hearing system into believing there's a real source at that angle. This occurs when the speakers differ in level by a certain amount.

The polarity of the two channels affects localization as well. To explain polarity, if the signals sent to two speaker channels are in polarity, they are in phase at all frequencies; both go positive in voltage at the same time. If the signals are out of polarity, they are 180° out of phase at all frequencies. One channel's signal goes positive when the other channel's signal goes negative. Opposite-polarity signals are sometimes incorrectly referred to as being *out of phase*.

If the signals are in opposite polarity between channels and equal level in both channels, the resulting image has a diffuse, directionless quality and cannot be localized. *Play CD track 4 to hear in-polarity and opposite-polarity signals.* If the signals are in opposite polarity and higher level in one channel than the other, the image often appears outside the bounds of the speaker pair. You'd hear an image left of the left speaker or right of the right speaker (Eargle, 1976).

Opposite polarity can occur in several ways. Two microphones are of opposite polarity if the wires to connector pins 2 and 3 are reversed in one microphone. Two speakers are of opposite polarity if the speaker-cable leads are reversed at one speaker. A single microphone might have different parts of its polar pattern in opposite polarity. For example, the rear lobe of a bidirectional pattern is opposite in polarity to the front lobe. If sound from a particular direction reaches the front lobe of the left-channel mic and the rear lobe of the right-channel mic, the two channels will be of opposite polarity. The resulting image of that sound source will either be diffuse or outside the speaker pair.

Requirements for Natural Imaging over Loudspeakers

To the extent that a sound recording and reproducing system can duplicate the interaural differences produced by a real source, the resultant image will be accurately localized. In other words, when reproduced sounds reaching a listener's ears have amplitude and phase differences corresponding to those of a real sound source at some particular angle, the listener perceives a well-fused, naturally focused image at that same angle. Conversely, when unnatural amplitude and phase relations are produced, the image appears relatively diffuse rather than sharp and is harder to localize accurately (Cooper and Bauck, 1980).

The required interaural differences for realistic imaging can be produced by certain interchannel differences. Placing an image in a precise location requires a particular amplitude difference versus frequency and phase difference versus frequency between channels. These have been calculated by Cooper and Bauck (1980) for several image angles. Gerzon (1980), Nakabayashi (1975), and Koshigoe and Takahashi (1976) have calculated the interaural or interchannel differences required to produce any image direction at a single frequency.

Figure A-8, for example, shows the interchannel differences required to place an image at 15° to the left of center when the speakers are placed ±30° in front of the listener (Cooper and Bauck, 1980).

As Figure A-8 shows, the interchannel differences required for natural imaging vary with frequency. Specifically, Cooper and Bauck (1980) indicate that interchannel amplitude differences are needed below approximately 1700 Hz and interchannel time differences are needed above that frequency (Cooper, 1987). Specifically:

- At low frequencies, the amplitude difference needed for a 15° image angle is about 10 dB.

- Between 1.7 and 5 kHz, the amplitude difference goes to approximately 0 dB. But there still is a phase difference to shift the image off-center.

- Above 1.7 kHz, the phase difference corresponds to a group delay (interchannel time difference) of about 0.547 millisecond, or 7.39 inches for a hypothetical spacing between microphones used for stereo recording.

189

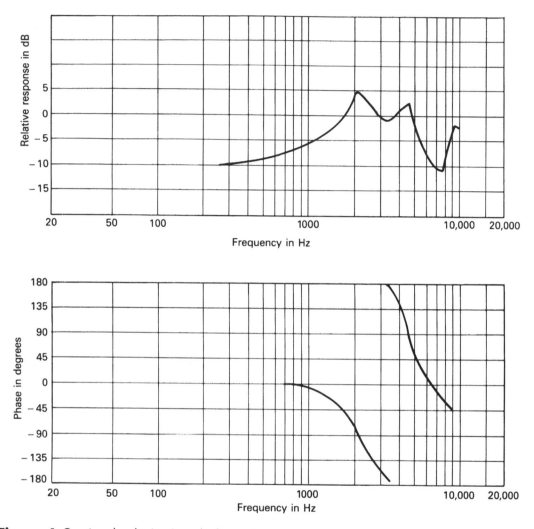

Figure A-8 Amplitude (top) and phase (bottom) of right channel relative to left channel, for image location 15° to the left of center when speakers are ±30° in front of listener.

This theory is based on the "shadowing" of sound traveling around a sphere. The description given here simplifies the complex requirements, but it conveys the basic idea. Cooper (1980) notes that "moderate deviations from these specifications might not lead to noticeable auditory distress or faulty imaging."

190

The Cooper–Bauck criteria can be met by recording with a dummy head whose signals are specially processed. A dummy head used for binaural recording is a modeled head with a flush-mounted microphone in each ear. Time and spectral differences between channels create the stereo images. (Spectral differences are amplitude differences that vary with frequency.)

Although a dummy-head binaural recording can provide excellent imaging over headphones, it produces poor localization over loudspeakers at low frequencies (Huggonet and Jouhaneau, 1987, Figure 13, p. 16) unless spatial equalization (a shuffler circuit) is used (Griesinger, 1989). Spatial equalization boosts the low frequencies in the difference ($L - R$) signal.

Binaural recording can produce images surrounding a listener wearing headphones but only frontal images over loudspeakers, unless a transaural converter is used. A transaural converter is an electronic device that converts binaural signals (for headphone playback) into stereo signals (for loudspeaker playback). Transaural stereo is a method of surround-sound reproduction using a dummy head for binaural recording, processed electronically to remove head-related crosstalk when the recording is heard over two loudspeakers (Bauer, 1961; Schroeder and Atal, 1963; Damaske, 1971; Eargle, 1976, pp. 122–123; Sakamoto et al., 1978, 1981, 1982; Mori et al., 1979; Cooper and Bauck, 1989; Moller, 1989).

Cooper recommends that, for natural imaging, the speakers' interchannel differences be controlled so that their signals sum at the ears to produce the correct interaural differences. According to Theile (1987), Cooper's theory (based on summing localization) is in error because it applies only to sine waves and may not apply to broadband spectral effects. He proposes a different theory of localization, the association model. This theory suggests that, when listening to two stereo loudspeakers, we ignore our interaural differences and instead use the speakers' interchannel differences to localize images.

The interchannel differences needed for best stereo, Theile says, are head related. The ideal stereo miking technique would use, perhaps, two ear-spaced microphones flush mounted in a head-size sphere and equalized for flat subjective response. This would produce interchannel spectral and time differences that, Theile claims, are optimum for stereo. The interchannel differences—time differences at low frequencies and amplitude differences at high frequencies—are the opposite of Cooper's requirements for natural stereo imaging. Time will tell which theory is closer to the truth.

Currently Used Image-Localization Mechanisms

The ear can be fooled into hearing reasonably sharp images between speakers by less sophisticated signal processing. Simple amplitude and/or time differences between channels, constant with frequency, can produce localizable images. Bartlett (1979, pp. 38, 40), Madsen (1957), Dutton (1962), Cabot (1977), Williams (1987), Blauert (1983), and Rumsey (1989) give test results showing image location as a function of interchannel amplitude or time differences. Bartlett's results are shown later in this appendix.

For example, given a speech signal, if the left channel is 7.5 dB louder than the right channel, an image will appear at approximately 15° to the left of center when the speakers are placed ±30° in front of the listener. A delay in the right channel of about 0.5 millisecond will accomplish the same thing, although image locations produced solely by time differences are relatively vague and hard to localize.

Griesinger notes that pure interchannel time differences do not localize low-pass-filtered male speech below 500 Hz over loudspeakers. Amplitude (level) differences are needed to localize low-frequency sounds. Either amplitude or time differences can localize high-frequency sounds (Griesinger, 1987).

With a baffled omni pair, the baffle attenuates or "shadows" high frequencies arriving from the opposite side, so the amplitude difference between mics increases with frequency. The time difference is constant with frequency.

The interchannel differences produced by coincident, near-coincident, and spaced-pair techniques are constant with frequency—just simple approximations of what is required. Still, reasonably sharp images are produced. Let's look at exactly how these differences localize images (Bartlett, 1979).

Localization by Amplitude Differences

The location of images between two loudspeakers depends in part on the signal amplitude differences between the loudspeakers. Suppose a speech signal is sent to two stereo loudspeakers, with the signal to each speaker identical except for an amplitude (level) difference (as shown in Figure A-9). We create an amplitude difference by inserting an attenuator in one channel.

Figure A-10 shows the approximate sound-image location between speakers versus the amplitude difference between channels, in decibels.

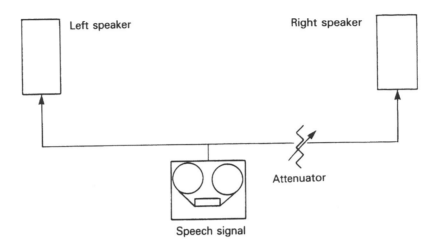

Figure A-9 Sending a speech signal to two stereo loudspeakers with attenuation in one channel.

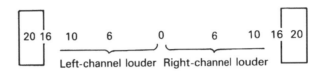

Figure A-10 Stereo-image location versus amplitude difference between channels, in dB (listener's perception).

A 0 dB difference (equal level from each speaker) makes the image of the sound source appear in the center, midway between the speakers. Increasing the difference places the image farther away from the center. A difference of 15–20 dB makes the image appear at only one speaker. *Play CD track 5 to hear a demonstration of image location versus level difference between channels.*

The information in this figure is based on carefully controlled listening tests. The data is the average of the responses of 10 trained listeners. They auditioned a pair of signal-aligned, high-quality loudspeakers several feet from the walls in a "typical" listening room, while sitting centered between the speakers at a 60° listening angle. Your own results may vary.

How can we create this effect with a stereo microphone array? Suppose two cardioid microphones are crossed at 90° to each other, with the grille of one microphone directly above the other (Figure A-11). The microphones

193

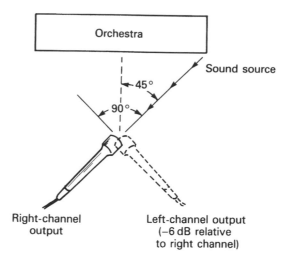

Figure A-11 Cardioids crossed at 90°, picking up a source at one end of an orchestra.

are angled 45° to the left and right of the center of the orchestra. Sounds arriving from the center of the orchestra will be picked up equally by both microphones. During playback, there will be equal levels from both speakers and, consequently, a center image is produced.

Suppose that the extreme right side of the orchestra is 45° off-center, from the viewpoint of the microphone pair. Sounds arriving from the extreme right side of the orchestra approach the right-aiming microphone on axis, but they will approach the left-aiming microphone at 90° off axis (as shown in Figure A-11). A cardioid polar pattern has a 6 dB lower level at 90° off axis than it has on axis. So, the extreme-right sound source will produce a 6 dB lower output from the left microphone than from the right microphone.

So we have a 6 dB amplitude difference between channels. According to Figure A-10, the image of the extreme-right side of the orchestra will now be reproduced right of center. Instruments in between the center and the right side of the orchestra will be reproduced somewhere between the 0 and 6 dB points.

If we angle the microphones farther apart, for example 135°, the difference produced between channels for the same source is around 10 dB. As a result, the right-side stereo image will appear farther to the right than it did with 90° angling. (Note that it is not necessary to aim the microphones exactly at the left and right sides of the ensemble.)

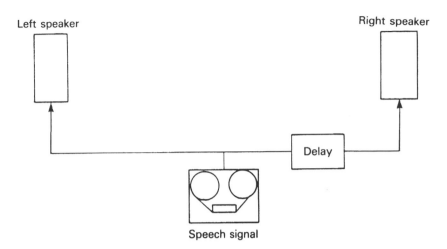

Figure A-12   Sending a speech signal to two speakers with one channel delayed.

The farther to one side a sound source is, the greater the amplitude difference between channels it produces and, thus, the farther from center is its reproduced sound image.

Localization by Time Differences

Phantom-image location also depends on the signal time differences between channels. Suppose we send the same speech signal to two speakers at equal levels but with one channel delayed (as in Figure A-12).

Figure A-13 shows the approximate sound-image location between speakers, with various time differences between channels, in milliseconds. A 0 millisecond difference (no time difference between speaker channels) makes the image appear in the center. As the time difference increases, the phantom image appears farther off-center. A 1.2–1.5 millisecond difference or delay is sufficient to place the image at only one speaker. *Play CD track 6 to hear a demonstration of image location versus time difference between channels.*

Spacing two microphones apart horizontally, even by a few inches, produces a time difference between channels for off-center sources. A sound arriving from the right side of the orchestra will reach the right microphone first, simply because it is closer to the sound source (as in Figure A-14). For example, if the sound source is 45° to the right, and the microphones are 8 inches apart, the time difference produced between

195

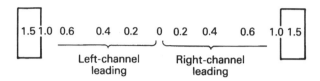

Figure A-13 Approximate image location versus time difference between channels, in milliseconds (listener's perception).

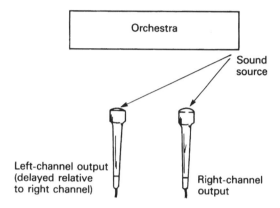

Figure A-14 Microphones spaced apart, picking up a source at one end of an orchestra.

channels for this source is about 0.4 millisecond. For the same source, a 20-inch spacing between microphones produces a 1.5 millisecond time difference between channels, placing the reproduced sound image at one speaker.

With spaced-pair microphones, the farther a sound source is from the center of the orchestra, the greater the time difference between channels and, thus, the farther from center is its reproduced sound image.

Localization by Amplitude and Time Differences

Phantom images also can be localized by a combination of amplitude and time differences. Suppose 90° angled cardioid microphones are spaced 8 inches apart (as in Figure A-15). A sound source 45° to the right will produce a 6 dB level difference between channels and a 0.4 millisecond

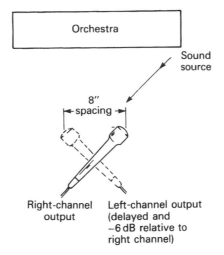

Figure A-15 Cardioids angled 90° and spaced 8 inches, picking up a source at one end of an orchestra.

difference between channels. The image shift of the 6 dB level difference adds to the image shift of the 0.4 millisecond difference to place the sound image at the right speaker. Certain other combinations of angling and spacing accomplish the same thing. *Play CD track 7 to hear a demonstration of image location versus level and time differences between channels.*

Summary

If a speech signal is recorded on two channels, its reproduced sound image will appear at only one speaker if—

- the signal is at least 15–20 dB lower in one channel;
- the signal is delayed at least 1.2–1.5 milliseconds in one channel; or
- the signal in one channel is lower in level and delayed by a certain amount.

When amplitude and time differences are combined to place images, the sharpest imaging occurs when the channel that is lower in level is also the channel that is delayed. If the higher level channel is delayed, image confusion results because of the conflicting time and amplitude cues.

We have seen that angling directional microphones (coincident placement) produces amplitude differences between channels. Spacing

197

microphones (spaced-pair placement) produces time differences between channels. Angling and spacing directional microphones (near-coincident placement) produces both amplitude and time differences between channels. These differences localize the reproduced sound image between a pair of loudspeakers.

Predicting Image Locations

Suppose you have a pair of microphones for stereo recording. Given their polar pattern, angling, and spacing, you can predict the interchannel amplitude and time differences for any sound-source angle. Hence, you can predict the localization of any stereo microphone array (in theory).

This prediction assumes that the microphones have ideal polar patterns, and that these patterns do not vary with frequency. It's an unrealistic assumption, but the prediction agrees well with listening tests.

The amplitude difference between channels in dB is given by

$$\Delta dB = 20 \log \left| \frac{a + b \cos (\theta_m / 2 - \theta_s)}{a + b \cos (\theta_m / 2 + \theta_s)} \right| \tag{A-1}$$

where

ΔdB = amplitude difference between channels, in dB;
$a + b \cos(\theta)$ = polar equation for the microphone;
Omnidirectional $a = 1$ $b = 0$
Bidirectional $a = 0$ $b = 1$
Cardioid $a = 0.5$ $b = 0.5$
Supercardioid $a = 0.366$ $b = 0.634$
Hypercardioid $a = 0.25$ $b = 0.75$
θ_m = angle between microphone axes, in degrees;
θ_s = source angle (how far off-center the sound source is), in degrees.

These variables are shown in Figure A-16.
The time difference between channels is given by

$$\Delta T = \frac{\sqrt{D^2 + [(S/2) + D \tan \theta_s]^2} - \sqrt{D^2 + [(S/2) - D \tan \theta_s]^2}}{C} \tag{A-2}$$

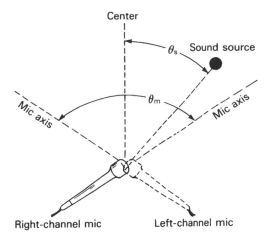

Figure A-16 Microphone angle (θ_m) and source angle (θ_s).

where

ΔT = time difference between channels, in seconds;
D = distance from the source to the line connecting the microphones, in feet;
S = spacing between microphones, in feet;
θ_s = source angle (how far off-center the sound source is), in degrees;
C = speed of sound (1130 feet per second).

These variables are shown in Figure A-17.

For near-coincident microphone spacing of a few inches, the equation can be simplified to this:

$$\Delta T = \frac{S \sin \theta_s}{C} \qquad \text{(A-3)}$$

where

ΔT = time difference between channels, in seconds;
S = microphone spacing, in inches;
θ_s = source angle, in degrees;
C = speed of sound (13,560 inches per second).

These variables are shown in Figure A-18.

Let's consider an example. If you angle two cardioid microphones 135° apart, and the source angle is 60° (as in Figure A-19), the dB difference produced between channels for that source is calculated as follows.

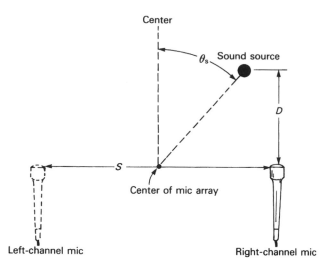

Figure A-17 Source angle (θ_s), mic-to-source distance (D), and mic spacing (S).

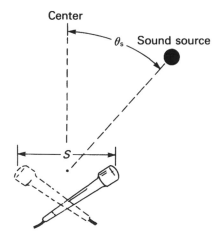

Figure A-18 Mic spacing (S) and source angle (θ_s).

For a cardioid, $a = 0.5$ and $b = 0.5$ (from the list following equation (A-1)). The angle between microphone axes, θ_m, is 135° and the source angle, θ_s, is 60°. That is, the sound source is 60° off-center. So the amplitude difference between channels, using equation (A-1), is

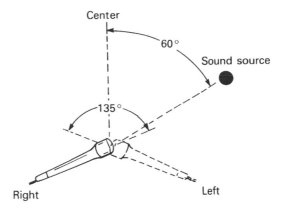

Figure A-19 Cardioids angled 135° apart, with a 60° source angle.

$$\Delta dB = 20 \log \left| \frac{0.5 + 0.5 \cos((135°/2) - 60°)}{0.5 + 0.5 \cos((135°/2) + 60°)} \right|$$

$$= 14\,dB \text{ amplitude difference between channels}$$

So, according to Figure A-10, that sound source will be reproduced nearly all the way at one speaker.

Here is another example. If you place two omnidirectional microphones 10 inches apart, and the sound source is 45° off-center, what is the time difference between channels? (Refer to Figure A-20.)

Microphone spacing, S, is 10 inches and source angle, θ_s, is 45°. By equation (A-3),

$$\Delta T = \frac{10 \sin 45°}{13,560}$$
$$= 0.52 \text{ millisecond time difference between channels}$$

So, according to Figure A-10, that sound source will be reproduced about halfway off-center.

Choosing Angling and Spacing

Many combinations of microphone angling and spacing are used to place the images of the ends of the orchestra at the right and the left speaker.

201

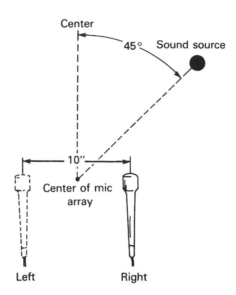

Figure A-20 Omnis spaced 10 inches apart, with a 45° source angle.

In other words, there are many ways to achieve a full stereo spread. You can use a narrow spacing and a wide angle, or a wide spacing and a narrow angle—whatever works. The particular angle and spacing you use is not sacred. Many do not realize this, and rely on a fixed angle and/or spacing, such as the ORTF (Office de Radiodiffusion Television Française) system (110°, 17 cm or 6.7 in). That is a good place to start, but if the reproduced stage width is too narrow, there's no harm in increasing the angle or spacing slightly.

If the center instruments are too loud, you can angle the mics farther apart while decreasing the spacing so that the reproduced stage width is unchanged. In this way, you can control the loudness of the center image to improve the balance.

To reduce pickup of early reflections from the stage floor and walls: (1) increase angling, (2) decrease spacing, and (3) place the mics closer to the ensemble. This works as follows:

1. Angling the mics farther apart softens the center instruments.
2. Decreasing the spacing between mics maintains the original reproduced stage width.
3. Since center instruments are quieter, you can place the mics closer to the ensemble and still achieve a good balance.

4. Since the mics are closer, the ratio of reflected sound to direct sound is decreased. You can add distant mics or artificial reverberation for the desired amount of hall ambience.

In general, a combination of angling and spacing (intensity and time differences) gives more accurate localization and sharper imaging than intensity or time differences alone (Griesinger, 1987).

Angling the mics farther apart increases the ratio of reverberation in the recording, which makes the orchestra sound farther away. Spacing the mics farther apart does not change the sense of distance, but it degrades the sharpness of the images.

Spaciousness and Spatial Equalization

The information in this section is from Griesinger's (1987) article, "New Perspectives on Coincident and Semi-coincident Microphone Arrays."

The spaciousness of a microphone array is the ratio of $L - R$ energy to $L + R$ energy in the reflected sound. Ideally, this ratio should be equal to or greater than 1. In other words, the sum and difference energy are equal. Spaciousness implies a low correlation between channels of the reflected sound.

Some microphone arrays with good spaciousness (a value of 1) are listed below:

- the spaced pair;
- the Blumlein pair (figure-eight mics crossed at 90°);
- the mid–side (MS) array with a cardioid mid pattern and a 1:1 M/S ratio;
- coincident hypercardioids angled 109° apart.

Spatial equalization or shuffling is a low-frequency shelving boost of difference ($L - R$) signals, and a complementary low-frequency shelving cut of sum ($L + R$) signals. This has two benefits:

1. It increases spaciousness, so that coincident and near-coincident arrays can sound as spacious as spaced arrays.
2. It aligns the low- and high-frequency components of the sound images, which results in sharper image focus.

You can build a spatial equalizer as shown in Griesinger's article. Or use an MS technique and boost the low frequencies in the $L - R$ or side signal, and cut the low frequencies in the $L + R$ or mid signal. The required boost or cut depends on the mic array, but a typical value is 4–6 dB shelving below 400 Hz. Excessive boost can split off-center images, with bass and treble at different positions. The correction should be done to the array before it is mixed with other mics.

Gerzon (1987) points out that the sum and difference channels should be phase compensated, as suggested by Vanderlyn (1957). Gerzon notes that spatial equalization is best applied to stereo microphone techniques not having a large antiphase reverberation component at low frequencies, such as coincident or near-coincident cardioids. With the Blumlein technique of crossed figure-eight mics, antiphase components tend to become excessive. He suggests a 2.4 dB cut in the sum ($L + R$) signal and a 5.6 dB boost in the difference ($L - R$) signal for better bass response.

Griesinger (1989) states,

Spatial equalization can be very helpful in coincident and semi-coincident techniques [especially when listening is done in small rooms]. Since the strongest localization information comes from the high frequencies, microphone patterns and angles can be chosen which give an accurate spread to the images at high frequencies. Spatial equalization can then be used to raise the spaciousness at low frequencies.

Alan Blumlein devised the first shuffler, revealed in his 1933 patent. He used it with two omni mic capsules spaced apart the width of a human head. The shuffler differenced the two channels (added them in opposite polarity). When two omnis are added in opposite polarity, the result is a single bidirectional pattern aiming left and right. Blumlein used this pattern as the side pattern in an MS pair (Lipshitz, 1990).

The frequency response of the synthesized bidirectional pattern is weak in the bass: it falls 6 dB/octave as frequency decreases. So Blumlein's shuffler circuit also included a 6 dB/octave low-frequency boost below 700 Hz to compensate.

The shuffler converts phase differences into intensity differences. The farther off-center the sound source is, the greater the phase difference between the spaced mics. And the greater the phase difference, the greater the intensity difference between channels created by the shuffler.

References

Bartlett, B. "Stereo Microphone Technique." *db*, Vol. 13, No. 12 (December 1979), pp. 34–46.

Bauer, B. "Stereophonic Earphones and Binaural Loudspeakers." *Journal of the Audio Engineering Society*, Vol. 9, No. 2 (April 1961), pp. 148–151.

Blauert, J. *Spatial Hearing*. Cambridge, MA: MIT Press, 1983.

Cabot, R. "Sound Localization in Two and Four Channel Systems: A Comparison of Phantom Image Prediction Equations and Experimental Data." Preprint No. 1295 (J3), Paper Presented at the *Audio Engineering Society 58th Convention*, November 4–7, 1977, New York.

Carterette, E. and Friedman, M. *Handbook of Perception. Vol. 4: Hearing*. New York: Academic Press, 1978.

Cooper, D. H. "Problems with Shadowless Stereo Theory: Asymptotic Spectral Status." *Journal of the Audio Engineering Society*, Vol. 35, No. 9 (September 1987), p. 638.

Cooper, D. and Bauck, J. "On Acoustical Specification of Natural Stereo Imaging." Preprint No. 1616 (X3), Paper Presented at the *Audio Engineering Society 65th Convention*, February 25–28, 1980, London.

Cooper, D. and Bauck, J. "Prospects for Transaural Recording." *Journal of the Audio Engineering Society*, Vol. 37, No. 1–2 (January–February 1989), pp. 9–19.

Damaske, P. "Head-Related Two-Channel Stereophony with Loudspeaker Reproduction." *Journal of the Acoustical Society of America*, Vol. 50, No. 4 (1971), pp. 1109–1115.

Dutton, G. "The Assessment of Two-Channel Stereophonic Reproduction Performance in Studio Monitor Rooms, Living Rooms, and Small Theatres." *Journal of the Audio Engineering Society*, Vol. 10, No. 2 (April 1962), pp. 98–105.

Eargle, J. *Sound Recording*. New York: Van Nostrand Reinhold Company, 1976.

Gardner, M. "Some Single and Multiple Source Localization Effects." *Journal of the Audio Engineering Society*, Vol. 21, No. 6 (July–August 1973), pp. 430–437.

Gardner, M. and Gardner, R. "Problems of Localization in the Median Plane-Effect of Pinnae Cavity Occlusion." *Journal of the Acoustical Society of America*, Vol. 53 (February 1973), pp. 400–408.

Gerzon, M. "Pictures of Two-Channel Directional Reproduction Systems." Preprint No. 1569 (A4), Paper Presented at the *Audio Engineering Society 65th Convention*, February 25–28, 1980, London.

Gerzon, M. Letter to the Editor, reply to comments on "Spaciousness and Localization in Listening Rooms and Their Effects on the Recording Technique." *Journal of the Audio Engineering Society*, Vol. 35, No. 12 (December 1987), pp. 1014–1019.

Griesinger, D. "New Perspectives on Coincident and Semi-coincident Microphone Arrays." Preprint No. 2464 (H4), Paper Presented at the *Audio Engineering Society 82nd Convention*, March 10–13, 1987, London.

Griesinger, D. "Equalization and Spatial Equalization of Dummy Head Recordings for Loudspeaker Reproduction." *Journal of the Audio Engineering Society*, Vol. 34, No. 1–2 (January–February 1989), pp. 20–29.

Harvey, F. and Schroeder, M. "Subjective Evaluation of Factors Affecting Two-Channel Stereophony." *Journal of the Audio Engineering Society*, Vol. 9, No. 1 (January 1961), pp. 19–28.

Huggonet, C. and Jouhaneau, J. "Comparative Spatial Transfer Function of Six Different Stereophonic Systems." Preprint No. 2465 (H5), Paper Presented at the *Audio Engineering Society 82nd Convention*, March 10–13, 1987, London.

Koshigoe, S. and Takahashi, S. "A Consideration on Sound Localization." Preprint No. 1132 (L9), Paper Presented at the *Audio Engineering Society 54th Convention*, May 4–7, 1976.

Lipshitz, S. Letter to the Editor. *Audio* (April 1990), p. 6.

Madsen, E. R. "The Application of Velocity Microphones to Stereophonic Recording." *Journal of the Audio Engineering Society*, Vol. 5, No. 2 (April 1957), p. 80.

Mehrgardt, S. and Mellert, V. "Transformation Characteristics of the External Human Ear." *Journal of the Acoustical Society of America*, Vol. 61, No. 6, (1977) p. 1567.

Moller, H. "Reproduction of Artificial-Head Recordings Through Loudspeakers." *Journal of the Audio Engineering Society*, Vol. 37, No. 1–2 (January–February 1989), pp. 30–33.

Mori, T., Fujiki, G., Takahashi, N., and Maruyama, F. "Precision Sound-Image-Localization Technique Utilizing Multi-track Tape Masters." *Journal of the Audio Engineering Society*, Vol. 27, No. 1–2 (January–February 1979), pp. 32–38.

Nakabayashi, K. "A Method of Analyzing the Quadraphonic Sound Field." *Journal of the Audio Engineering Society*, Vol. 23, No. 3 (April 1975), pp. 187–193.

Rumsey, F. *Stereo Sound for Television*. Boston: Focal Press, 1989.

Sakamoto, N., Gotoh, T., Kogure, T., and Shimbo, M. "On the Advanced Stereophonic Reproducing System 'Ambience Stereo.'" Preprint No. 1361 (G3), Paper Presented at the *Audio Engineering Society 60th Convention*, May 2–5, 1978, Los Angeles.

Sakamoto, N., Gotoh, T., Kogure, T., and Shimbo, M. "Controlling Sound-Image Localization in Stereophonic Reproduction, Part I." *Journal of the Audio Engineering Society*, Vol. 29, No. 11 (November 1981), pp. 794–799.

Sakamoto, N., Gotoh, T., Kogure, T., and Shimbo, M. "Controlling Sound-Image Localization in Stereophonic Reproduction, Part II." *Journal of the Audio Engineering Society*, Vol. 30, No. 10 (October 1982), pp. 719–722.

Schroeder, M. and Atal, B. "Computer Simulation of Sound Transmission in Rooms." *IEEE Convention Record*, Part 7 (1963), pp. 150–155.

Shaw, E. "Transformation of Sound Pressure Levels from the Free Field to the Eardrum in the Horizontal Plane." *Journal of the Acoustical Society of America*, Vol. 56, No. 6 (December 1974), pp. 1848–1861.

Theile, G. "On the Stereophonic Imaging of Natural Spatial Perspective via Loudspeakers: Theory." In *Perception of Reproduced Sound 1987*, eds. Soren Bech and O. Juhl Pedersen. Munich: Institut fur Rundfunktechnik, 1987.

Vanderlyn, P. British Patent 781,186 (August 14, 1957).

Vanderlyn, P. "Auditory Cues in Stereophony." *Wireless World* (September 1979), pp. 55–60.

Wallach, H., Newman, E., and Rozenzweig, M. "The Precedence Effect in Sound Localization." *Journal of the Audio Engineering Society*, Vol. 21, No. 10 (December 1973), pp. 817–826.

Williams, M. "Unified Theory of Microphone Systems for Stereophonic Sound Recording." Preprint No. 2466 (H6), Paper Presented at the *Audio Engineering Society 82nd Convention*, March 10–13, 1987, London.

SPECIFIC FREE-FIELD STEREO MICROPHONE TECHNIQUES

Some stereo microphone techniques work better than others. Each method has different effects. A few techniques provide sharper imaging; some create a narrow stage effect; some have exaggerated separation, and so on. In this appendix, I compare the characteristics of several specific stereo microphone techniques. All of these use free-field microphones; the next appendix covers stereo techniques using boundary microphones and dummy heads.

Localization Accuracy

One characteristic that varies among different types of arrays is localization accuracy. Localization is accurate if instruments at the sides of the ensemble are reproduced from the left or right speaker; instruments halfway off-center are reproduced halfway between the center and one speaker, and so on. In other words, there is little or no distortion of the geometry of the musical ensemble.

For example, suppose your stereo speakers are spaced the same distance apart as you're sitting from them, so that each speaker is ±30° off-center. (This is the recommended arrangement for good stereo.) If the orchestral width "seen" by the microphone pair is 90°, we want sources that are 45° to one side of center to be reproduced out of only one speaker. Sources 22.5° off-center should be reproduced halfway between the center of the speaker pair and one speaker (15° off-center).

Figure B-1 illustrates this. In Figure B-1(a), the letters A through E represent live sound-source positions relative to the microphone pair. In Figure B-1(b), the corresponding images of these sources are accurately localized between the speaker pair.

Spacing or angling the microphones more than is necessary to achieve a full stereo spread produces an "exaggerated separation" effect: Instruments near the center are reproduced to the extreme left or right, rather than slightly off-center. Instruments exactly in the center are still reproduced

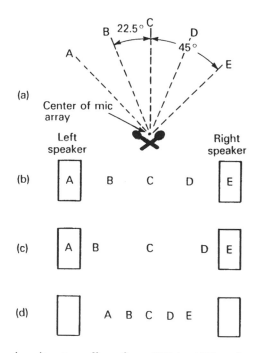

Figure B-1 Stereo localization effects for a 90° (±45°) orchestral width: (a) letters A through E represent live sound-source positions (top view); (b) accurately localized images between speakers (listener's perception); (c) exaggerated separation effect; and (d) narrow stage effect.

between the speakers (see Figure B-1(c)). Conversely, too little angling or spacing gives a poor stereo spread or a "narrow stage" effect (see Figure B-1(d)). *Play CD track 20 to hear exaggerated separation and play CD tracks 9–10 to hear the narrow stage effect.*

A listening test was performed to determine the localization accuracy of various stereo microphone techniques, for a 90° orchestral width (Bartlett, 1979). Recordings were made of a speech source at 0°, 22.5°, and 45° relative to the microphone pair (as in Figure B-2(a)). Tests were made in an anechoic chamber and in a reverberant gymnasium. Listeners were asked to note the reproduced sound-image locations for several techniques. The image locations of the anechoic and reverberant recording rooms were averaged, with results shown in Figure B-2(b).

Since results may vary under different listening conditions, this information is meant to be indicative, rather than definitive. Different listeners hear stereo effects differently, so your perceptions may not agree exactly with those shown. Still, Figure B-2 lets you compare one technique to another. *Play CD tracks 8–24 to hear imaging comparisons of various stereo mic techniques.*

The 90° orchestral width used is arbitrary. The actual width of the orchestra varies with the size of the ensemble and the mic-to-source distance. If the orchestral width is more than 90°, the stereo spread of all these techniques is wider than shown in Figure B-2(b).

The closer to the ensemble a microphone array is placed, the greater is the orchestral width as seen by the microphone pair, and, thus, the wider is the stereo spread (up to the limit of the speaker spacing).

Examples of Coincident-Pair Techniques

In general, coincident cardioids tend to give a narrow stereo spread and lack a sense of air or spaciousness. Imaging at high frequencies is not optimum because there is no time difference between channels, which, according to Cooper, is essential. Also, when microphones are angled apart, they receive much of the sound off axis. Many microphones have off-axis coloration (a different frequency response on and off axis).

Coincident techniques are mono-compatible: the frequency response is the same in mono and stereo. That is because there are no phase or time differences between channels to cause phase cancellations if the two channels are mixed to mono.

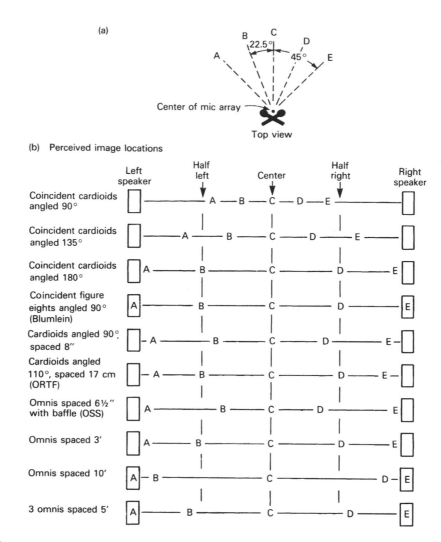

Figure B-2 Image location of some stereo mic arrays versus source position: (a) letters A through E are live speech-source positions relative to the mic array; (b) images A through E are the perceived image locations that each stereo mic array produces.

Coincident Cardioids Angled 180° Apart

According to Figure B-2(b), it seems reasonable to angle two coincident cardioid microphones 180° apart to achieve maximum stereo spread (as shown in Figure B-3). However, sounds arriving from straight ahead approach each

212

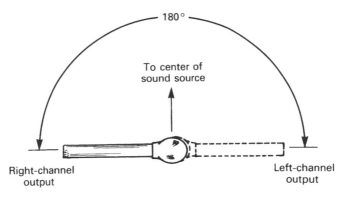

Figure B-3   Coincident cardioids angled 180° apart.

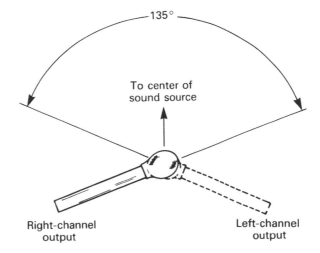

Figure B-4   Coincident cardioids angled 135° apart.

microphone 90° off axis. The 90° off-axis frequency response of some micro-phones is weak in high frequencies, giving a dull sound to instruments in the center of the orchestra. In addition, it has been the experience of another experimenter, Michael Gerzon (1976, p. 36), that 180° angling places the reproduced reverberation to the extreme left and right. *Play CD track 11 to hear the imaging of coincident cardioids angled 180° apart.*

Coincident Cardioids Angled 120°–135° Apart

A 120°–135° angle between microphones might be a better compromise. Gerzon has reported that the 120° angle gives a uniform spread of rever-beration between speakers, while the 135° angle (Figure B-4) provides a

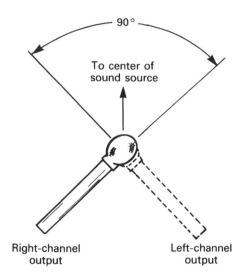

Figure B-5 Coincident cardioids angled 90° apart.

slightly wider stereo spread. These angles are useful when you don't want the reproduced ensemble to spread all the way between speakers. For a wider stereo spread, you can use a near-coincident or spaced pair. However, the 135° angle just described can provide a full stereo spread if the orchestral width or source angle is 150°. *Play CD track 10 to hear the stereo imaging of coincident cardioids angled 120° apart.*

Coincident Cardioids Angled 90° Apart

Angling cardioids at 90° (Figure B-5) reproduces most of the reverberation in the center. It gives a narrow stage width, unless the ensemble surrounds the microphone pair in a semicircle (180° source angle). *Play CD track 9 to hear the stereo imaging of coincident cardioids angled 90° apart.*

Blumlein Technique

This classic method uses two coincident bidirectional mics angled 90° apart (Figure B-6). As shown in Figure B-2(b), it provides accurate localization. According to Gerzon (1976) and the listening tests, it also provides sharp imaging, a fine sense of depth, and the most uniform possible spread of reverberation across the reproduced stereo stage. It has the sharpest perceived image focus of any system, other than spatially equalized systems (Huggonet and Jouhaneau, 1987, p. 11, Figure 8).

214

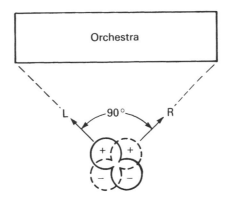

Figure B-6 The Blumlein or stereosonic technique (coincident bidirectionals crossed at 90°).

Note that each bidirectional pattern has a rear lobe in opposite polarity to the front lobe. If a sound source is more than 45° off-center (say, off to the left side), it is picked up by the front-left lobe and the back-right lobe. These are opposite in polarity. This creates antiphase information between channels, which produces vague localization. For this reason, the microphones should aim at the extreme-left and -right ends of the performing ensemble. This prevents sound sources from being outside the 45° limit. However, this limitation fixes the mic-to-source distance. You can't adjust this distance to vary the sense of perspective, unless you also change the angle between microphones or the size of the musical ensemble.

Another drawback is that the microphones pick up a large amount of reverberation. If you place the microphone pair closer to the ensemble to increase the direct/reverb ratio, the stereo spread becomes excessive and instruments in the center of the ensemble are emphasized. In addition, instruments at either end of the ensemble are reproduced with opposite-polarity signals from both channels, so they are not localized.

The Blumlein technique works best in a wide room with minimal side-wall reflections, where strong signals are not presented to the sides of the stereo pair (Streicher and Dooley, 1985).

Hypercardioids Angled 110° Apart

Shown in Figure B-7, this method give accurate localization. Listening tests also reveal sharp imaging and very good spaciousness. This array has the widest in-phase region of any array that has a spaciousness of 1 (Griesinger, 1987). The tight pattern of the hypercardioid allows a more

215

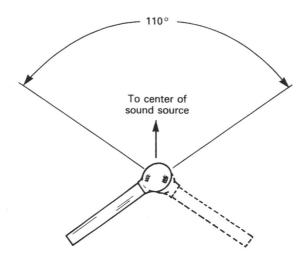

Figure B-7 Hypercardioids angled 110° apart.

distant placement than with crossed cardioids. As for drawbacks, hyper-cardioid microphones tend to have a bass roll-off; but this can be corrected with equalization (bass boost).

Another coincident technique is the mid–side (MS) technique, which will be covered in detail later in this appendix.

Examples of Near-Coincident-Pair Techniques

If you start with a coincident pair and space the mics a few inches apart (making them near coincident), the stereo spread will increase. So will the spaciousness and depth, because of the random-phase relationships (low correlation) between channels at high frequencies.

Near-coincident methods are not mono-compatible: if both channels are combined to mono, there are dips in the frequency response caused by phase cancellations. Also, since the microphones are angled apart, the sound source might be reproduced with off-axis coloration.

The ORTF and DIN Systems

The listening tests summarized in Figure B-2(b) reveal that the 110° angled, 17-cm (6.7-inch) spaced cardioid array (the ORTF system) and the 90° angled, 8-inch (20-cm) spaced cardioid array (the DIN system) tend to

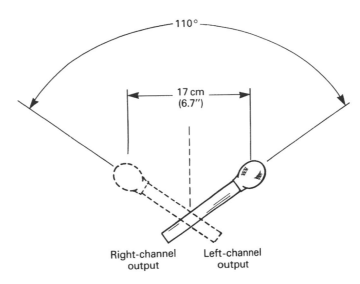

Figure B-8 The ORTF system: cardioids angled 110° and spaced 17 cm (6.7 inches) apart.

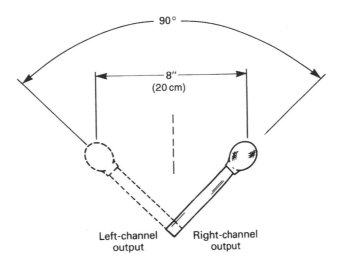

Figure B-9 The DIN system: cardioids angled 90° and spaced 20 cm (8 inches) apart.

provide accurate localization. These two methods are shown in Figures B-8 and B-9. According to a listening test conducted by Carl Ceoen (1972), the ORTF system was preferred over several other stereo miking techniques. It provided the best overall compromise of localization accuracy, image

217

sharpness, an even balance across the stage, and ambient warmth. *Play CD track 13 to hear the stereo imaging of the ORTF technique.*

The origin of the ORTF system was described by Condamines (1978). The 17 cm (6.7 in) spacing was chosen because it provided the best image stability with head motion, assuming a speaker angle of ±30°. The 110° angle was chosen because it provided the best image precision and placement when used with a 17 cm (6.7 in) spacing. Condamines reported that, if the mic angle is less than 110°, the sound stage usually does not spread all the way between speakers; if the angle is greater than 110°, the center image becomes weak (a hole-in-the-middle effect).

The ORTF image position varies with frequency, according to calculation (Bernfeld and Smith, 1978) and perception (Huggonet and Jouhaneau, 1987, p. 14, Figure 11).

The NOS System

Shown in Figure B-10, this system was proposed by the Dutch Broadcasting Foundation. It uses two cardioids angled 90° apart and spaced 30 cm (11.8 inches) horizontally. Since the spacing of the NOS system exceeds the 90° angled, 8-inch-spaced array in the listening test, we could expect it to

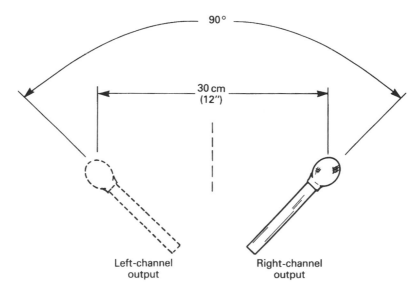

Figure B-10 The NOS system: cardioids angled 90° and spaced 30 cm (11.8 inches) apart.

have a slightly wider stereo spread for halfway-left and -right instruments. *Play CD track 14 to hear the stereo imaging of the NOS technique.*

Examples of Spaced-Pair Techniques

In general, listeners commented that the spaced-pair methods give relatively vague, hard-to-localize images for off-center sources. These methods are useful when you want diffuse images for special effect. Spaced arrays have a pleasing sense of spaciousness. This is produced artificially by the random-phase relationships between channels, and by opposite-polarity signals at various frequencies (Lipshitz, 1986).

Spaced-pair techniques are not mono-compatible: peaks and dips in the frequency response of the direct sound occur when both channels are combined to mono. This effect may or may not be audible, because reverberation approaches the microphones from all angles, and each angle of sound incidence relates to a different pattern of phase cancellations. The reverberation randomizes the frequencies of these cancellations, so that the effect is less audible.

An advantage of the spaced-pair technique is that it allows the use of omnidirectional condenser microphones, which have a more extended low-frequency response than directional microphones. That is, the tone quality is warmer and fuller in the bass. Of course, you can equalize directional microphones to have flat bass response at a distance.

Another advantage is that the listening area for good stereo is wider than with coincident-pair techniques. The spaced-pair delay cues counteract the amplitude imbalance that occurs when the listener sits off-center.

Many instruments, such as the flute, have nulls in their sound-radiation pattern that vary with the note played. Thus, one mic of a spaced pair might pick up a note at a low signal level, while the other mic would pick it up at a high signal level, so the image would wander with the note played. However, one mic will pick up notes that the other mic misses. Our ears have the same ability due to their spacing. Thus, the spaced pair offers the potential for better fidelity (no missed notes) at the expense of wandering images (Lemon, 1989).

You can use cardioids or other unidirectional patterns in a spaced array to reduce pickup of hall reverberation. These patterns, however, tend to have less bass than omnis. Spaced figure-eight mics have very little off-axis coloration.

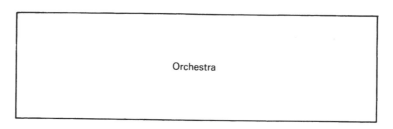

Orchestra

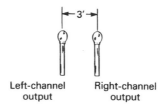

Left-channel output   Right-channel output

Figure B-11   Omnis spaced 3 feet apart.

Omnis Spaced 3 Feet Apart

Shown in Figure B-11, this method gives fairly accurate localization (Figure B-2(b)), but with poorly focused imaging of off-center sources. A 2-foot spacing would give more accurate localization (*as in CD track 19*). Since omnis must be placed relatively close to a performing ensemble for an acceptable direct/reverb ratio, this array is likely to overemphasize the center instruments. That is, the microphone pair is most sensitive to instruments in the center of the orchestra, with reduced pickup of the sides.

Telarc often uses a 2-foot spaced pair, angled 90° to each other, about 10 feet high, plus a pair of flanking omnis spaced 10–15 feet each side of center. The flanking mics are 2–3 dB below the center pair. The center mics are panned partly left and right; the flanks are panned hard left and right.

Omnis Spaced 10 Feet Apart

Shown in Figure B-12, this spacing provides a more even coverage of the orchestra (a better balance). However, spacings greater than 3 feet give an exaggerated separation effect, in which instruments slightly off-center are reproduced full-left or -right (Figure B-2(b)). This dispels the myth that spaced microphones should be as far apart as the playback loudspeakers. Instruments directly in the center of the ensemble are still reproduced

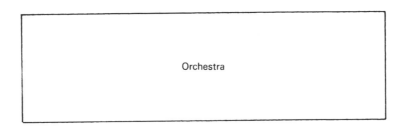

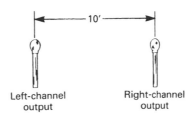

Figure B-12   Omnis spaced 10 feet apart.

exactly between the speakers. *CD track 10 demonstrates the stereo imaging of two mics spaced 6 feet apart.*

Three Omnis Spaced 5 Feet Apart (10 Feet End to End)

With this method (Figure B-13), a third microphone is placed between the other two, mixed in at an approximate equal level, and split to both channels. This reduces stereo separation while maintaining full coverage of the orchestra (see Figure B-2). The three-spaced-omnis technique is often used by Telarc Records. Image focus and mono-compatibility are fair to good.

Decca Tree

Developed in 1954 by the Decca Record Company, the Decca Tree is an array of three spaced omnidirectional mics (Figure B-14; Gayford, 1994; www.josephson.com/deccatree). Mic spacing depends on the desired amount of width and spaciousness. The center mic is placed slightly forward of the outer pair. Because the center mic's signal precedes that of the outer pair, the center mic helps to "solidify" the center image.

As for placement, the triangle of mics is mounted about 10–12 feet above the stage, just behind the conductor. The outer pair is angled outward

221

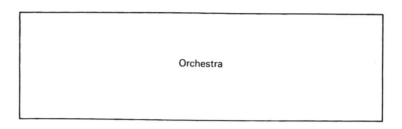

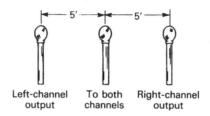

Left-channel output   To both channels   Right-channel output

Figure B-13 Three omnis spaced 5 feet apart.

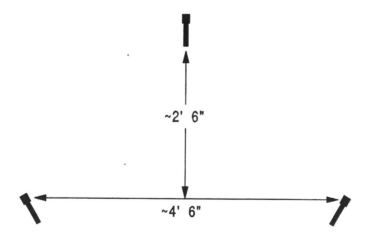

Figure B-14 Decca Tree stereo microphone technique.

to point at the edges of the stage, so that the edges are picked up with the best high-frequency response. Central sounds are on axis to the center microphone. The center mic may exacerbate the comb-filtering effects that occasionally occur with spaced pairs.

Sometimes, an additional pair of flanking mics is used near the edges of the orchestra or about one-third of the way in. These flanking mics face diagonally across the orchestra and help to add width and spaciousness. All mics are mixed at an equal level. The center mic is panned to center, both left mics are panned hard left, and both right mics are panned hard right.

Mic spacing varies with the venue and the ensemble size. The center mic or the outriggers might be omitted in some cases.

Examples of Baffled-Omni Techniques

Sphere Microphone, SASS-P MKII

These mics can be called either "baffled-omni mics" or "boundary mics." They are described in Appendix C, Stereo Boundary-Microphone Arrays, under the headings "The Stereo Ambient Sampling System (SASS)" and "Sphere Microphones." *Play CD track 17 to hear the stereo imaging of a sphere microphone.*

Optimal Stereo Signal or Jecklin Disk

The Jecklin disk uses two omnidirectional microphones spaced 16.5 cm (6.5 inches) apart and separated by a disk with a diameter of 28 cm (11⅞ inches) (Jecklin, 1981). The disk is hard and is covered with flat, absorbent material to reduce reflections (Figure B-15). The Schneider disk is the same but is covered with two foam hemispheres. The optimal stereo signal (OSS) system could be called *quasi binaural*, in that the human binaural hearing system also uses two omni "microphones" separated by a baffle (the head).

Below 200 Hz, both microphones receive the same amplitude, and the array acts like closely spaced omnis. As frequency increases, the disk becomes more of a sound barrier, which makes the array increasingly directional. At high frequencies, the array acts like a near-coincident pair of subcardioids angled 180° apart.

Since both channels receive the same signal level at low frequencies, stereo localization at low frequencies can be due only to the capsule spacing, which causes direction-dependent delays. But, according to Griesinger (1987), delay panning does not create localizable images below 500 Hz. If that is true, the OSS system localizes only above 200 Hz.

According to the inventor (Jecklin, 1981), "the stereo image is nearly spectacular, and the sound is rich, full, and clear." It "seems to be superior to all other recording methods." The full sound is probably due to the use

223

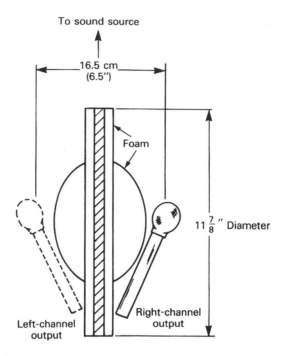

Figure B-15   The OSS system or Jecklin disk. Omnis spaced 16.5 cm (6.5 inches) apart and separated by a foam-covered disk of 28 cm (11⅞ inches) diameter.

of omnidirectional condenser microphones, which have an extended low-frequency response. *Play CD track 16 to hear the stereo imaging and full bass of the Jecklin Disk method.*

Listening tests (Figure B-2(b)) show that the OSS stereo spread for a 90° orchestral width is somewhat narrow. But, since the system uses omni microphones, it is usually placed close to the ensemble, where the angular width of the ensemble is wide. This results in a wider stereo spread.

Other Coincident-Pair Techniques

Let's return to coincident-pair methods and go over some specific techniques in detail.

Mid–Side

This method uses a middle (mid) microphone capsule aiming straight ahead toward the center of the performing ensemble, plus a side-aiming

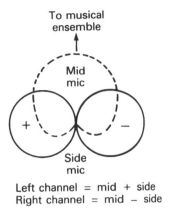

Left channel = mid + side
Right channel = mid − side

Figure B-16 MS stereo microphone technique.

(side) bidirectional microphone capsule. These capsules are coincident and at right angles to each other (as shown in Figure B-16). The mid capsule is most commonly cardioid, but it can be any pattern.

The outputs of both capsules are summed (mixed) to produce the left-channel signal and are differenced (mixed in opposite polarity) to produce the right-channel signal. In effect, this creates two virtual polar patterns angled apart:

$$M + S = L$$
$$M - S = R$$

For example, suppose that the mid capsule is omnidirectional and the side capsule is bidirectional. Also suppose that the sensitivity of both capsules is set equal. When you add these two patterns together, you get a cardioid aiming 90° to the left. When you subtract these patterns (add them in opposite polarity) you get a cardioid aiming 90° to the right. Thus, an MS microphone with an omni mid capsule is equivalent to two coincident cardioids angled 180° apart. An MS microphone with a bidirectional mid capsule is equivalent to two figure-eight mics crossed at 90° (the Blumlein technique).

Some stereo microphones have switchable polar patterns. Changing the mid-capsule pattern changes the pattern and angling of the virtual polar patterns. The more directional the mid mic is, the more directional

225

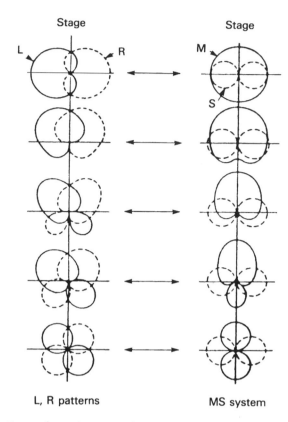

Figure B-17 Equivalent directional patterns for MS system, with mid pattern varied. (From a letter by Les Stuck to *db* magazine, March 1981.)

are the sum-and-difference (virtual) polar patterns (Figure B-17). Consequently, you can change the apparent distance from the sound source by changing the mid pattern.

MS Matrix Box

The *M* and *S* outputs of the microphone are connected to an MS matrix box or decoder. This decoder uses either a tapped transformer or an active circuit to sum-and-difference the *M* and *S* signals. The output of the box is a left- and a right-channel signal. Some sources of MS matrix decoders are given in Chapter 12 in the section "MS Matrix Decoders."

A rotating knob in the box controls the ratio of the mid signal to the side signal. By varying the ratio of mid-to-side signals, you change the

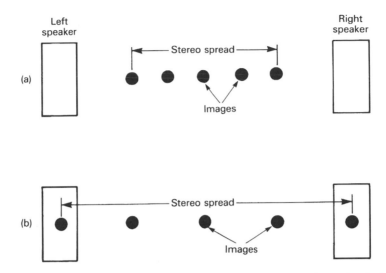

Figure B-18 Effects of varying *M/S* ratio on stereo spread: (a) high *M/S* ratio gives a narrow spread and (b) low *M/S* ratio gives a wide spread.

polar pattern and angling of the left and right virtual mic capsules. In turn, this varies the stereo spread and the ratio of direct-to-reverberant sound. As you turn up the side signal, the stereo-spread widens and the ambience increases, as shown in Figure B-18. The optimum starting *M/S* ratio is near 1:1.

A dual-mode matrix lets you vary the spread of an MS array, and also vary the spread of any standard left/right stereo mic array.

In Chapter 9 under the heading "Stereo-Spread Control," I describe how to use a computer digital audio workstation (DAW) to act as an MS matrix.

Some MS mics have an MS matrix built in. These microphones have a left-right output and usually include a stereo-spread switch in the mic body.

MS Advantages

A major advantage of the MS system is that you can control the stereo spread from a remote location, or after the recording is done. This feature is especially useful for live concerts, where you can't change the microphone array during the concert. Since the stereo spread is adjustable, the MS system can be made to have accurate localization.

If you record the *M* and *S* signals directly to a two-track recorder during the concert, you can play them back through a matrix decoder after the concert and adjust the stereo spread then. In postproduction,

227

you can vary the spread from very narrow (mono) to very wide. While recording the concert, you monitor the outputs of the matrix decoder but do not record them.

The MS method has another advantage: it is fully mono-compatible. If you sum the left and right channels to mono, you get just the output of the forward-facing mid capsule. This is shown in the following equations:

$$\text{Left} = (M + S)$$

$$\text{Right} = (M - S)$$

$$\text{Left} + \text{Right} = (M + S) + (M - S) = 2M$$

With XY- or near-coincident techniques, the center image is formed by adding the outputs of two angled directional capsules. If they are not perfectly matched in frequency and phase responses, the fusion of the center image can be degraded. But the MS system has very sharp center imaging because the center image is the output of the single mid capsule.

MS Disadvantages

The MS system has been criticized for a lack of warmth, intimacy, and spaciousness (Ceoen, 1972; Griesinger, 1987). However, Griesinger states that MS recordings can be made more spacious by giving the low frequencies a shelving boost of 4 dB (starting with +2 dB at 600 Hz) in the $L - R$ or side signal, with a complementary shelving cut in the $L + R$ or mid signal.

There are other disadvantages to the MS technique. It requires a matrix decoder, which is extra hardware to take on location. A final disadvantage is that the stereo spread and direct-to-reverb ratio are interdependent: you can't change one without changing the other.

When the signals from an MS stereo microphone are mixed to mono, the resulting signal is only from the front-facing mid capsule. If this capsule's pattern is cardioid, sound sources to the far left or right will be attenuated. Thus, the balance might be different in stereo and mono. If this is a problem, use an XY-coincident pair rather than MS.

Double MS Technique

Skip Pizzi recommends a double MS technique, which uses a close MS microphone mixed with a distant MS microphone. One MS microphone is close to the performing ensemble for clarity and sharp imaging, and the other is 50–75 feet out in the hall for ambience and depth. The distant mic could be replaced by an XY pair for lower cost (Pizzi, 1984).

For a comprehensive discussion of the MS system, see Streicher and Dooley (1985).

SoundField Microphone

This British microphone (shown in Figure B-19) is an elaboration on the MS system. It uses four closely spaced cardioid mic capsules arranged in a tetrahedron and aiming outward. Their outputs are phase shifted to make the capsules seem perfectly coincident.

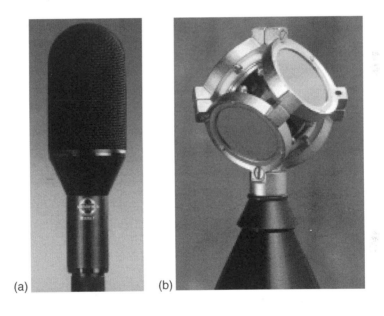

(a)   (b)

(c)

Figure B-19 SoundField Mk V Microphone: (a) external view and (b) internal view, showing capsules.

229

The capsule outputs are called the *A-format signals*. They are electronically matrixed by a control unit to produce:

- An omnidirectional component (the sound pressure).
- A vertical pressure-gradient component.
- A left-right pressure-gradient component.
- A fore-aft pressure-gradient component.

These B-format signals can be further processed into stereo or surround signals. With a remote-control box, the user can adjust polar patterns, azimuth (horizontal rotation), elevation (vertical tilt), dominance (apparent distance), and angle (stereo spread) (Farrar, 1979; Streicher and Dooley, 1985).

As for drawbacks, the microphone system is expensive and requires a complex matrix circuit. But it is the world's premier microphone for spatial recording. Several models are described in Chapter 12 under the heading "Surround Microphones."

Coincident Systems with Spatial Equalization (Shuffler Circuit)

Coincident-pair systems have been criticized for a lack of spaciousness. However, as discovered by Blumlein (1958), Vanderlyn (1954), and Griesinger (1986, 1987), the focus and spaciousness can be improved by a shuffler circuit (spatial equalization). This circuit decreases stereo separation at high frequencies or increases separation at low frequencies, in order to align the image locations at low and high frequencies. To increase low-frequency separation, the circuit applies a shelving boost to low frequencies in the left-minus-right (difference) signal, and applies a complementary cut to the left-plus-right (sum) signal.

Griesinger reports that spatially equalized coincident or near-coincident arrays have very sharp imaging, and sound as spacious as a spaced array. As stated earlier, MS recordings can be made more spacious by boosting the bass $+4$ dB shelving ($+2$ dB at 600 Hz) in the $L - R$ or side signal, and cutting the sum signal by the same amount.

Other Near-Coincident-Pair Techniques

Let's look more closely at some unusual near-coincident miking methods.

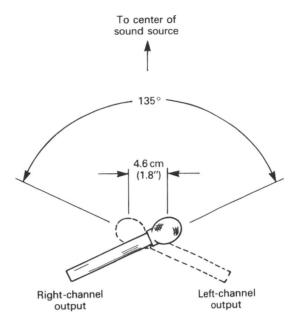

Figure B-20   Stereo 180 system: hypercardioids angled 135° and spaced 4.6 cm (1.8 inches) apart.

Stereo 180 System

Another near-coincident method is the Stereo 180 System developed by Olson (1979), shown in Figure B-20. It uses two hypercardioid pattern microphones angled 135° apart, and spaced 4.6 cm (1.8 inches) horizontally. The hypercardioid patterns have opposite-polarity rear lobes, which create the illusion that the reproduced reverberation is coming from the sides of the listening room as well as between the speakers. The localization accuracy and image focus of the array are reported to be very good.

Faulkner Phased-Array System

Invented by Tony Faulkner (1982), this method uses two bidirectional (figure-eight) microphones aiming straight ahead with axes parallel and spaced 20 cm (7.87 inches) apart (Figure B-21). The plane of maximum sound-path difference coincides with the null in the directional polar pattern of the microphones. Since the microphones are aimed forward rather than angled apart, you can place them farther from the ensemble

231

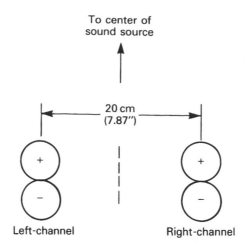

Figure B-21 Faulkner phased-array system: two figure eights spaced 20 cm (7.8 inches) apart.

for a better balance. This distant placement also lets you place the microphones at ear height, rather than raised. Faulkner says that the array is not mono-compatible in theory but has presented no problems in practice.

Sometimes Faulkner adds a pair of omnidirectional microphones 2–3 feet apart, flanking the figure eights. These omnis add ambient spaciousness.

Near-Coincident/Spaced-Pair Hybrid

John Eargle, Director of Recording at Delos International Inc., prefers to use a combination of near-coincident and spaced-pair methods (*Symphonic Sound Stage* CD). A quasi-ORTF pair is placed about 4 feet behind the conductor, 9–12 feet high. This pair is flanked by two omnis 12–16 feet apart, typically 6 dB below the main pair. The ORTF pair provides sharp imaging and depth, while the spaced pair adds width to the strings and time cues from the hall. Since the spaced pair uses omnidirectional mics, low-frequency reproduction is excellent.

A second stereo pair is placed up to 30 feet behind the main pair to capture hall reverb. The woodwinds often are picked up with an overhead pair, and accent mics are added if necessary for soloists, harp, celeste, and other instruments.

Comparisons of Various Techniques

Many studies have been done comparing standard stereo miking techniques. The results of some of these are presented here. They do not all agree. *Play CD tracks 8–24 to compare various techniques for yourself.*

Michael Williams, "Unified Theory of Microphone Systems for Stereophonic Sound Recording" (1987)

Michael Williams calculated the recording angle and standard deviation of several fixed systems. *Recording angle* means the angle subtended by the sound source required for a speaker-to-speaker stereo spread. It is the angular width of the performing ensemble (as seen by the microphone array) that causes a full stereo spread.

The *standard deviation* represents geometric distortion of the sound stage. The bigger the standard deviation in degrees, the wider is the image separation of halfway-left and -right instruments. If standard deviation is 0°, instruments halfway left in the orchestra are reproduced halfway left between the loudspeakers (that is, at 15° off-center for speakers separated ±30°). If standard deviation is large, this is the exaggerated separation effect mentioned earlier.

Here are his findings for various fixed mic arrays:

Coincident cardioids at 90°

The recording angle is ±90° (180° in all). In other words, the orchestra must form a semicircle (180°) around the microphone pair to be reproduced from speaker to speaker.

The standard deviation is about 6°. In other words, an instrument that is half-right in the orchestra would be reproduced 6° beyond half-right.

Coincident figure eights at 90° (Blumlein)

Recording angle is ±45° (90° in all).

Standard deviation is about 5°.

Cardioids angled 110° and spaced 17 cm (6.7 in) (ORTF)

Recording angle is ±50° (100° in all).

Standard deviation is about 5°.

Cardioids angled 90° and spaced 30 cm (11.8 in) (NOS)

Recording angle is ±40° (80° in all).

Standard deviation is about 4°.

Omnis spaced 50 cm (20 inches)

Recording angle is ±50° (100° in all).

Standard deviation is about 8°.

Williams's article has graphs showing the calculated recording angle and standard deviation for a wide range of polar patterns, anglings, and spacings, as well as other useful information.

Carl Ceoen, "Comparative Stereophonic Listening Tests" (1972)

Carl Ceoen used listening tests to compare several typical stereo techniques. He reported the following average resolution distortion (image focus or sharpness) for these methods:

XY (coincident cardioids angled 135°): 3°

MS (equivalent to coincident hypercardioids angled apart): 5.5°

Blumlein (coincident bidirectionals angled 90°): 4°

ORTF (coincident cardioids at 110°, 17 cm (6.7 in)): 3°

NOS (cardioids angled 90° and spaced 30 cm (11.8 in)): 4°

Pan pot: 3°

According to Ceoen, the listening audience agreed that the ORTF system was the best overall compromise, and that the MS system lacked intimacy.

Benjamin Bernfeld and Bennett Smith, "Computer-Aided Model of Stereophonic Systems" (1978)

Bernfeld and Smith computed the image location versus frequency for various stereo miking techniques. The better the coincidence of image locations at various frequencies, the sharper the imaging. Here are the condensed results:

Blumlein (coincident bidirectionals at 90°): Image focus is good except near the speakers; there, high frequencies are reproduced with a wider stereo spread than low frequencies.

Coincident cardioids angled 90° apart: Image focus is very good, but the stereo spread is very narrow.

Coincident cardioids angled 120° apart: Image focus is fairly good, but the stereo spread is narrow.

Coincident hypercardioids angled 120° apart: Image focus is good, but not excellent because high frequencies around 3 kHz are reproduced with a wider spread than low frequencies.

Coincident hypercardioids angled 120° apart, compensated with Vanderlyn's shuffler circuit (Vanderlyn, 1954): Excellent image focus and stereo spread.

Blumlein (coincident bidirectionals at 90°), compensated with shuffler circuit: Very good image focus and stereo spread.

ORTF (cardioids angled 110° and spaced 17 cm (6.7 in)): Good image focus; low frequencies have narrow spread and high frequencies have wide spread.

ORTF with hypercardioids: Similar to the above, with wider stereo separation.

Two omnis spaced 9.5 feet: Poor image focus; high frequencies have much wider spread than low frequencies; exaggerated separation effect.

Three cardioids spaced 5 feet: Poor image focus as above, with exaggerated separation at high frequencies.

C. Huggonet and J. Jouhaneau, "Comparative Spatial Transfer Function of Six Different Stereophonic Systems" (1987)

Huggonet and Jouhaneau used a modulated tone burst at various frequencies, plus a violin, with listening tests to compare the spatial transfer function of six different stereophonic systems. Each system has an angular dispersion (image spread) that depends on frequency. In general, the angular dispersion of coincident systems was least. The Blumlein array gave the sharpest imaging, the dummy head, and the NOS system the worst. The dummy head gave the best depth perception, followed by ORTF. MS gave the worst depth perception.

235

M. Hibbing, "XY and MS Microphone Techniques in Comparison" (1989)

In comparing XY and MS coincident methods, Hibbing concluded that MS has several advantages over XY:

1. The MS system can use an omnidirectional mid element, but the XY system cannot use omnidirectional capsules. Since an omni capsule generally has better low-frequency response than a uni, the MS system can have better low-frequency response than the XY system.

2. With MS, any stereo spread can be had with any polar pattern. XY is more limited.

3. With MS, a wider source angle is usable than with XY if polar patterns with a low bidirectional component are used.

4. With MS, the mid element aims at the center of the sound source, so most of the sound arrives close to on axis. With XY, most of the sound arrives off axis and is subject to off-axis coloration.

5. With MS, both the mid- and side-polar patterns are more uniform with frequency than the patterns in the XY configuration. Consequently, the left/right polar patterns generated by MS are more uniform with frequency than those of XY.

6. With MS, the stereo spread is easy to control by a fader. With XY, the stereo spread must be adjusted mechanically. MS allows stereo-spread adjustment after the session; XY does not.

7. With MS, the mid (mono sum) signal is independent of the stereo spread, so it stays consistent and predictable. With XY, the mono sum varies with the angle between the microphones.

Wieslaw Woszczyk, "A New Method for Spatial Enhancement in Stereo and Surround Recording" (1990)

Using female speech and a soprano recorder as a sound source, Woszcyk recorded the source with several stereo arrays in the diffuse field (29–80 feet from the source). Blind listening tests were done using a stereo pair of speakers and a Dolby surround system. The latter system used front-left and -right speakers, a center front speaker, a center rear speaker, and left/right surround speakers.

The results are briefly summarized below: in general, listening in Dolby surround reduces the stereo separation (stage width) because of the center speaker. Mic techniques for Dolby surround should be optimized to counteract this effect.

In these descriptions, *stage width* means the perceived width of the stage (about ±65° in front of the mic pair). *Spatial effect* means the perceived spaciousness of the concert hall:

XY at 90°: Very narrow stage width, narrow spatial effect.

XY at 180°: Extremely wide stage width and weak center image in stereo, but fairly accurate in surround. This method gave the best spatial effect of the listening test: wide, intense, and natural.

ORTF with cardioids: Fairly accurate stage width in stereo but much narrower in surround. Narrow spatial effect.

ORTF with hypercardioids: Wide stage width, "split" spatial effect.

Blumlein: Accurate stage width up to ±45° in stereo, slightly narrower in Dolby surround. Wide and smooth spatial effect.

14-inch-spaced omni pair: Somewhat narrow stage width in stereo, even less in surround. Smooth and natural spatial effect.

Dummy head: Wide stage width in stereo, narrower in surround. Superior spatial effect: wide and smooth.

Pressure Zone Microphone (PZM) wedge (two 18-inch × 29-inch hard baffles angled 45°): Overly wide stage width in stereo but accurate in surround. Superior, natural spatial effect. Somewhat honky coloration. Spherical microphones should produce equally good imaging but without coloration.

Summary

Although these experimenters disagree in certain areas, they all agree that widely spaced microphones give poorly focused imaging and that the Blumlein technique gives sharp imaging. Blumlein and Bernfeld say that the imaging of the Blumlein array can be further sharpened with a shuffler or spatial equalizer. Ceoen's results indicate that ORTF is best, but others report less-than-optimal image focus with ORTF.

The most accurate systems for frontal stereo appear to be coincident or near-coincident arrays with spatial equalization, or dual MS arrays. The near-coincident/spaced-pair hybrid method used by Delos also works quite well.

It helps to know about all the stereo techniques in order to conquer the acoustic problems of various halls or to create specific effects. No particular technique is magic; you often can improve the results by changing the microphone angling or spacing.

I recommend the following recording, which demonstrates the imaging differences among various free-field stereo microphone techniques: *The Performance Recordings Demonstration of Stereo Microphone Technique* (PR-6-CD), recorded by James Boyk, Mark Fischman, Greg Jensen, and Bruce Miller; available at www.performancerecordings.com/albums.html.

References

The Website www.dpamicrophones.com has a section called "Microphone University," which includes an excellent discussion of stereo microphone techniques.

Bartlett, B. "Stereo Microphone Technique." *db*, Vol. 13, No. 12 (December 1979), pp. 310–346.

Bernfeld, B. and Smith, B. "Computer-Aided Model of Stereophonic Systems." Paper Presented at the *Audio Engineering Society Preprint No. 1321, 59th Convention*, 1978–2002, p. 14.

Blumlein, A. "British Patent Specification." *Journal of the Audio Engineering Society*, Vol. 6, No. 2 (April 1958), p. 91. Also in *Stereophonic Techniques Anthology*. New York: Audio Engineering Society, 1986.

Ceoen, C. "Comparative Stereophonic Listening Tests." *Journal of the Audio Engineering Society*, Vol. 20, No. 1 (January–February 1972), pp. 19–27. Also in *Stereophonic Techniques Anthology*. New York: Audio Engineering Society, 1986.

Condamines, R. "La Prise de Son." In *Stereophonic*. Paris and New York: Masson Publishers, 1978.

Farrar, K. "Sound Field Microphone." *Wireless World* (October 1979).

Faulkner, T. "Phased Array Recording." *The Audio Amateur* (January 1982).

Gayford, M. *Microphone Engineering Handbook*. Trowbridge, Wiltshire: Focal Press, 1994.

Gerzon, M. "Blumlein Stereo Microphone Technique." *Journal of the Audio Engineering Society*, Vol. 24, No. 11 (January–February 1976), p. 36.

Griesinger, D. "Spaciousness and Localization in Listening Rooms and Their Effects on Recording Technique." *Journal of the Audio Engineering Society*, Vol. 34, No. 4 (April 1986), pp. 255–268.

Griesinger, D. "New Perspectives on Coincident and Semi-coincident Microphone Arrays." Preprint No. 2464 (H4), Paper Presented at the *Audio Engineering Society 82nd Convention*, March 10–13, 1987, London.

Hibbing, M. "XY and MS Microphone Techniques in Comparison." *Journal of the Audio Engineering Society*, Vol. 37, No. 10 (October 1989), pp. 823–831.

Huggonet, C. and Jouhaneau, J. "Comparative Spatial Transfer Function of Six Different Stereophonic Systems." Preprint No. 2465 (H5), Paper Presented at the *Audio Engineering Society 82nd Convention,* March 10–13, 1987, London.

Jecklin, J. "A Different Way to Record Classical Music." *Journal of the Audio Engineering Society,* Vol. 29, No. 5 (May 1981), pp. 329–332. Also in *Stereophonic Techniques Anthology.* New York: Audio Engineering Society, 1986.

Lemon, J. "Spacing for Fidelity," letter to the editor. *Recording Engineer/Producer* (September 1989), p. 76.

Lipshitz, S. "Stereo Microphone Techniques: Are the Purists Wrong?" *Journal of the Audio Engineering Society,* Vol. 34, No. 9 (September 1986), pp. 716–744.

Olson, L. "The Stereo-180 Microphone System." *Journal of the Audio Engineering Society,* Vol. 27, No. 3 (March 1979), pp. 158–163. Also in *Stereophonic Techniques Anthology.* New York: Audio Engineering Society, 1986.

Pizzi, S. "Stereo Microphone Techniques for Broadcast." Preprint No. 2146 (D-3), Paper Presented at the *Audio Engineering Society 76th Convention,* October 8–11, 1984, New York.

Streicher, R. and Dooley, W. "Basic Stereo Microphone Perspectives—A Review." *Journal of the Audio Engineering Society,* Vol. 33, No. 7–8 (July–August 1985), pp. 548–556. Also in *Stereophonic Techniques Anthology.* New York: Audio Engineering Society, 1986.

The Symphonic Sound Stage, Vol. 2. Delos Compact Disc D/CD 3504.

Vanderlyn, P. British Patent Specification 23989 (1954).

Williams, M. "Unified Theory of Microphone Systems for Stereophonic Sound Recording." Preprint No. 2466 (H-6), Paper Presented at the *Audio Engineering Society 82nd Convention,* March 10–13, 1987, London.

Woszcyk, W. "A New Method for Spatial Enhancement in Stereo and Surround Recording." Preprint No. 2946, Paper Presented at the *Audio Engineering Society 89th Convention,* September 21–25, 1990, Los Angeles.

C

STEREO BOUNDARY-MICROPHONE ARRAYS

Boundary microphones (discussed in Chapter 7) can make excellent stereo recordings. This appendix explains the characteristics of several boundary-mic arrays.

First, here are some ways to create basic stereo arrays using boundary microphones (Bartlett, 1999):

- To make a spaced-pair boundary array, space two boundary microphones a few feet apart. Place them on the floor, on a wall, or on stand-mounted panels.
- To make a coincident array, mount two boundary mics back-to-back on a 2-foot × 2-foot clear plastic panel, with the edge of the panel aiming at the sound source.
- To make a near-coincident array, mount each boundary microphone on a separate panel, and angle the panels apart to form a "V." Or use two directional boundary mics on the floor, angled and spaced.

Boundary microphones can be placed directly on the floor or can be raised above the floor. We explain several stereo techniques using both methods.

Techniques Using Floor-Mounted Mics

You can place two boundary microphones on the floor to record in stereo. Floor mounting provides several advantages:

- Phase cancellations due to sound reflections off the floor are eliminated.
- Floor mounting provides the best low-frequency response for boundary microphones.
- The mics are very easy to place.
- The mics are nearly invisible. At live concerts, hiding the microphones is often the main consideration.

When a floor-placed boundary array is used to record an orchestra, the front-row musicians are usually reproduced too loudly, due to their relative proximity to the microphones. Musical groups with little front-to-back depth—such as small chamber groups, jazz groups, or soloists—may be the best application for this system.

Let's consider specific techniques for floor-mounted microphones.

Floor-Mounted Boundary Microphones Spaced 4 Feet Apart

Listening tests showed that a spacing of 3–4 feet between microphones is sufficient for a full stereo spread, when the sides of the musical ensemble are 45° off-center, from the viewpoint of the center of the microphone array (see Figure C-1). *Listen to CD track 22 to hear the stereo imaging of two floor-mounted PZMs (Pressure Zone Microphones) placed 3 feet apart.*

With a floor-placed array, the stereo spread decreases as the sound-source height increases. For example, if you record a group of people who are standing, the spread will be narrower than if the people are sitting. That's because the higher the source is, the less the time difference is between microphones.

Spaced boundary mics have the same drawbacks as spaced free-field mics: poorly focused images, potential lack of mono-compatibility, and large phase differences between channels. *Listen to CD track 22 to hear the stereo imaging of two floor-mounted PZMs placed 3 feet apart.*

Two advantages, however, are a warm sense of ambience and a good stereo effect even for off-center listeners. And, with a spaced pair, you can use omnidirectional boundary microphones without Plexiglas boundaries. So the low-frequency response is excellent and the mics are inconspicuous.

242

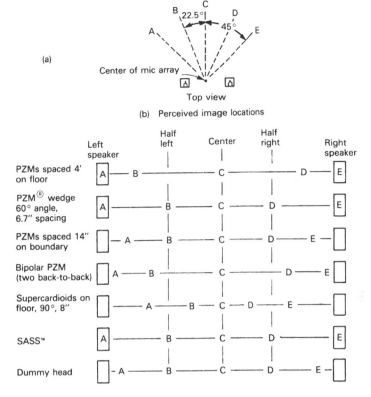

Figure C-1 Stereo localization of some stereo boundary-microphone arrays: (a) letters A through E are live speech-source positions relative to mic array; (b) A through E are the perceived image locations produced by the mic arrays.

Floor-Mounted Directional Boundary Microphones

Two of these can be set up as a near-coincident pair or a spaced pair. For near-coincident use, place the mics on the floor side by side and angle them apart (Figure C-2). Adjust their angling and spacing for the desired stereo spread. This is an effective arrangement for recording stage plays or musicals. Other mics are needed for the pit orchestra.

As shown in Figure C-1, a floor-mounted array of supercardioid boundary mics, angled 90° and spaced 8 inches, provides a narrow stage width. More spacing or angling is needed for accurate localization. The image focus is sharper with this arrangement than with spaced boundary microphones. *CD track 23 demonstrates the stereo imaging of a pair of cardioid boundary mics angled 90° and spaced 12 inches.*

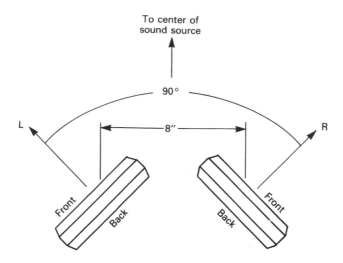

Figure C-2 Floor-mounted directional boundary microphones set in a near-coincident array.

Optimal Stereo Signal Boundary-Microphone Floor Array

In this configuration by Josephson Engineering (1988), two boundary microphones are on opposite sides of a hard, absorbent baffle or Jecklin disk cut in half. This array has the characteristics of the Optimal Stereo Signal (OSS) system described in the previous appendix plus the advantages of boundary miking.

"The Musician's Ear" Stereo Boundary Microphone

This unit has two condenser mics flush-mounted in opposite sides of a wooden baffle that sits on the floor (www.performancerecordings.com/ear.html).

Floor-Mounted Boundary Microphones Configured for Mid–Side

The mid–side (MS) technique can be applied to boundary microphones. The following method was invented by Jerry Bruck (1985) of Posthorn Recordings. The mid unit is an omnidirectional boundary microphone; the side unit is a small-diameter bidirectional condenser microphone mounted a few millimeters (a few thousandths of an inch) above the omni unit.

The bidirectional microphone is close enough to the boundary to prevent phase cancellations between direct and reflected sounds over most of the audible spectrum.

Since the mid microphone is a boundary type, it has the same high-frequency response anywhere around it (no off-axis coloration). This contributes to very sharp stereo imaging. And since the mid capsule is an omni condenser unit, it has excellent low-frequency response. The system is low profile and unobtrusive.

Like other floor-mounted methods, this system is limited to recording small ensembles or soloists. It also could be used on a piano lid. No microphones are made this way; you must set a bidirectional microphone over a boundary microphone to form the array.

Techniques Using Raised Boundary Mics

If you mount two omni mic capsules on boundaries, such as panels or a sphere, the mics become directional. You can raise the panel or sphere several feet off the floor to record large ensembles in stereo. We will describe two mics of this type.

The Stereo Ambient Sampling System

This is a stereo condenser microphone using boundary-microphone technology (Bartlett, 1989; Billingsley, 1987, 1989a, 1989b, 1989c). It is designed to give sharp stereo imaging for loudspeaker or headphone reproduction. The device is a mono-compatible, near-coincident array. According to soundmixer Gary Pillon, a stereo ambient sampling system (SASS) mounted on a Steadicam platform gives an excellent match between audio and video perspectives.

The SASS uses two PZMs mounted on boundaries to make each microphone directional (as shown in Figure C-3).

For each channel, an omni mic capsule is mounted very near a 5-inch square boundary. The two boundaries are angled left and right of center. The sound diffraction of each boundary, along with a foam barrier between the capsules, creates a directional polar pattern at high frequencies. The patterns aim left and right of center, much like a near-coincident array. The capsules are "ear-spaced" 17 cm (6.7 inches) apart.

245

Figure C-3 Crown SASS-P MKII PZM stereo microphone (courtesy: Crown International, Inc.).

The polar patterns of the boundaries and the spacing between capsules have been chosen to provide natural perceived stereo imaging. Like an artificial head (described in Appendix D), the SASS localizes images by time and spectral differences between channels.

The foam barrier or baffle between the capsules limits acoustic crosstalk between the two sides at higher frequencies. Although the microphone capsules are spaced apart, little phase cancellation occurs when both channels are combined to mono because of the shadowing effect of the baffle. That is, despite phase differences between channels, the extreme amplitude differences (caused by the baffle) reduce phase cancellations in mono.

Application notes for the SASS are given in Bartlett (1989) and Billingsley (1987, 1989a, 1989b, 1989c, 1990).

Sphere Microphone

A sphere microphone uses a hard globe 8 inches in diameter, with a pair of pressure-response omni mic capsules flush-mounted in either side, 180° apart. Two examples are the Neumann KFM 100 and the Schoeps KFM 6.

Like the SASS, a spherical stereo mic uses time and spectral differences between channels to create stereo images. A circuit corrects the frequency response and phase response of the capsules in the sphere.

Claimed benefits are accurate and sharp imaging, excellent reproduction of depth, extended low-frequency response, and low pickup of wind and vibration. The sphere shape is used because it provides the least diffraction (disturbance of the sound field). As a result, the frequency response is flat not only for sounds in front of the sphere but also for reverberant, diffuse sound. *Play CD track 17 to hear the stereo imaging of a sphere microphone. It is two mini omni mics taped to either side of a head-size sphere.*

The mic is largely mono-compatible because, in the bass frequencies, phase shift is small; and, in the high frequencies, the acoustic shadow of the sphere produces strong interchannel differences making phase cancellations less probable.

A sphere mic is not the same as a dummy head. Sphere-mic recordings are for speaker listening; dummy-head recordings are for headphone listening.

References

Bartlett, B. "An Improved Stereo Microphone Array Using Boundary Technology: Theoretical Aspects." Preprint No. 2788 (A-1), Paper Presented at the *Audio Engineering Society 86th Convention*, March 7–10, 1989, Hamburg.

Bartlett, B. *Crown Boundary Microphone Application Guide*. Elkhart, IN: Crown International, 1999. www.crownaudio.com

Billingsley, M. US Patent 4,658,931 (April 21, 1987).

Billingsley, M. "Practical Field Recording Applications for an Improved Stereo Microphone Array Using Boundary Technology." Preprint No. 2788 (A-1), Paper Presented at the *Audio Engineering Society 86th Convention*, March 7–10, 1989a, Hamburg.

Billingsley, M. "An Improved Stereo Microphone Array for Pop Music Recording." Preprint No. 2791 (A-2), Paper Presented at the *Audio Engineering Society 86th Convention*, March 7–10, 1989b, Hamburg.

Billingsley, M. "A Stereo Microphone for Contemporary Recording." *Recording Engineer/Producer* (November 1989c).

Billingsley, M. "Theory and Application of a New Near-Coincident Stereo Microphone Array for Soundtrack, Special Effects and Ambience." Paper

Presented at the *Audio Engineering Society 89th Convention*, September 21–25, 1990, Los Angeles.

Bruck, J. "The Boundary Layer Mid/Side (M/S) Microphone: A New Tool." Preprint No. 2313 (C-11), Paper Presented at the *Audio Engineering Society 79th Convention*, October 12–16, 1985, New York.

Josephson Engineering. *Catalog*. San Jose, CA: Josephson Engineering, 1988.

D

BINAURAL TECHNIQUES

This appendix covers binaural recording with an artificial (dummy) head. The head contains a microphone flush-mounted in each ear. You record with these microphones and play back the recording over headphones. This process can re-create the locations of the original performers and their acoustic environment with exciting realism.

You can substitute your own head for the artificial head by placing miniature condenser microphones in your ears, or on your temples, and recording with them. Some podcasts are made this way.

Thanks to the popularity of MP3 players with earphones, many people have the opportunity to hear binaural recordings.

Binaural Recording and the Artificial Head

Binaural (two-ear) recording starts with an artificial head or dummy head. This is a model of a human head with a flush-mounted microphone in each ear (Figure D-1). These microphones capture the sound arriving at each ear. The microphones' signals are recorded. When this recording is played back over headphones, your ears hear the signals that originally appeared at the dummy head's ears (Figure D-2). That is, the original sound at each ear is reproduced (Geil, 1979; Genuit and Bray, 1989; Peus, 1989; Sunier, 1989a, 1989b, 1989c).

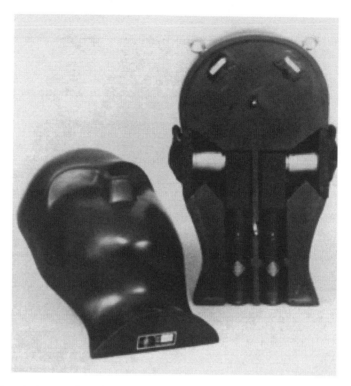

Figure D-1   A dummy head used for binaural recording (courtesy: Neumann USA).

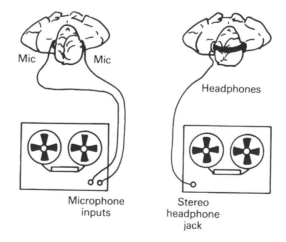

Figure D-2   Binaural recording and headphone playback.

Binaural recording works on the following premise. When we listen to a natural sound source in any direction, the input to our ears is just two one-dimensional signals: the sound pressures at the eardrums. If we can re-create the same pressures at the listener's eardrums as would have occurred "live," we can reproduce the original listening experience, including directional information and reverberation (Moller, 1989).

Binaural recording with headphone playback is the most spatially accurate method now known. The re-creation of sound-source locations and room ambience is startling. Often, sounds can be reproduced all around your head—in front, behind, above, below, and so on. You may be fooled into thinking that you're hearing a real instrument playing in your listening room. A catalog and demos of binaural recordings are available from The Binaural Source at www.binaural.com, headed by binaural expert John Sunier.

As for drawbacks: the artificial head is conspicuous, which limits its use for recording live concerts; it is not mono-compatible; and it is relatively expensive. Some sources for dummy heads are given in Chapter 12 under the heading "Dummy Heads and Headworn Binaural Mics."

How It Works

An artificial head picks up sound as a human head does. The head is an obstacle to sound waves at middle to high frequencies. On the side of the head away from the sound source, the ear is in a sonic shadow: the head blocks high frequencies. In contrast, on the side of the head toward the source, there is a pressure buildup (a rise in the frequency response) at middle to high frequencies.

The folds in the pinna (outer ear) also affect the frequency response by reflecting sounds into the ear canal. These reflections combine with the direct sound, causing phase cancellations (dips in the response) at certain frequencies.

The human eardrum is inside the ear canal, which is a resonant tube. The ear canal's resonance does not change with sound-source direction, so the ear canal supplies no localization cues. For this reason, it is omitted in most artificial heads. Typically, the microphone diaphragm is mounted nearly flush with the head, 4 mm (0.16 in) inside the ear canal.

To summarize: The head and outer ear cause peaks and dips in the frequency response of the sound received. These peaks and dips vary with the angle of sound incidence; they vary with the sound-source location. The frequency response of an artificial head is different in different

251

directions. In short, the head and outer ear act as a direction-dependent equalizer.

Each ear picks up a different spectrum of amplitude and phase because one ear is shadowed by the head and the ears are spaced apart. These interaural differences vary with the source location around the head.

When the signals from the dummy-head microphones are reproduced over headphones, you hear the same interaural differences that the dummy head picked up. This creates the illusion of images located where the original sources were.

Physically, an artificial head is a near-coincident array using boundary microphones: the head is the boundary, and the microphones are flush-mounted in this boundary. The head and outer ears create directional patterns that vary with frequency. The head spaces the microphones about 6½ inches apart. Some dummy heads include shoulders or a torso, which aids front/back localization in human listening but can degrade it in binaural recording and playback (Griesinger, 1989).

The microphones in a near-coincident array are directional at all frequencies and use no baffle between them. In contrast, the mics in an artificial head are omni at low frequencies and unidirectional at high frequencies (due to the head baffle effect).

Ideally, the artificial head is as solid as a human head, to attenuate sound passing through it (Sunier, 1989c). For example, the Head Acoustics artificial head is made of molded, dense fiberglass (Genuit and Bray, 1989). In contrast, the Sonic Studios GUY and LiteGUY artificial heads are made of absorbent Sorbothane.

As we said, you can substitute your own head for the artificial head by placing miniature condenser microphones in your ears and recording with them. The more that a dummy head and ears are shaped like your particular head and ears, the better the reproduced imaging. Thus, if you record binaurally with your own head, you might experience more precise imaging than you would if you recorded with a dummy head. This recording will have a nonflat response because of head diffraction (which I will explain later).

Core Sound (www.core-sound.com) is the world's largest manufacturer of binaural microphones. The company offers miniature omni condenser mics that can be clipped onto eyeglass earpieces. These mics make excellent binaural recordings. Sonic Studios (www.sonicstudios.com) has a similar product, DSM (Dimensional Stereo Microphones) that are worn on the temples rather than in the ear. Based on the head related transfer function (HRTF), DSM mics are said to provide better stereo over loud-speakers than binaural mics can provide. An explanation of HRTF was given at the beginning of Chapter 12.

Another substitute for a dummy head is a head-size sphere with flush-mounted microphones where the ears would be. This system, called the *Kugelflachenmikrofon*, was developed by Gunther Theile for improved imaging over loudspeakers (Griesinger, 1989). See Appendix C under the heading "Sphere Microphones." *CD track 17 demonstrates the stereo imaging of a sphere microphone. Listen to it over headphones as well as loudspeakers.*

Some commercial products are listed in Chapter 12 under the headings "Dummy Heads and Headworn Binaural Mics" and "Stereo Microphones."

In-Head Localization

You might hear the binaural images inside your head, rather than outside. One reason has to do with head movements. When you listen to a sound source that is outside your head and move your head slightly, you hear small changes in the arrival-time differences at your ears. This is a cue to the brain that the source is outside your head. Small movements of your head help to externalize sound sources. But the dummy head lacks this cue because it is stationary.

Another reason for in-head localization is that the conch resonance of the pinna is disturbed by most headphones. The conch is the large cavity in the pinna just outside the ear canal. If you equalize the head-phone signal to restore the conch resonance, you hear images outside the head (Cooper and Bauck, 1989).

Artificial-Head Equalization

An artificial head (or a human head) has a nonflat frequency response due to the head's diffraction, the disturbance of a sound field by an obstacle. The diffraction of the head and pinnae creates a very rough frequency response, generally with a big peak around 3 kHz for frontal sounds. Therefore, binaural recordings sound tonally colored unless custom equalization is used. Some artificial heads have built-in equalization that compensates for the effect of the head.

What is the best equalization for an artificial head to make it sound tonally like a conventional flat-response microphone? Several equalization schemes have been proposed:

- *Diffuse-field equalization*: This compensates for the head's average response to sounds arriving from all directions (such as reverberation in a concert hall).

- *Frontal free-field equalization*: This compensates for the head's response to a sound source directly in front, in anechoic conditions.

- *10° averaged, free-field equalization*: This compensates for the head's response to a sound source in anechoic conditions, averaged over ±10° off-center.

- *Free field with source at ±30° equalization*: This compensates for the head's response to a sound source 30° off-center, in anechoic conditions. This is a typical stereo loudspeaker location.

The Neumann KU-100 and KEMAR artificial heads use diffuse-field equalization, which Theile also recommends. However, Griesinger (1989) found that the Neumann head needed additional equalization to sound like a Calrec Soundfield microphone: approximately +7 dB at 3 kHz and +4 dB at 15 kHz. He prefers either this equalization or a 10° averaged free-field response for artificial heads. The Head Acoustics head, developed by Gierlich and Genuit, is equalized flat for free-field sounds in front (Genuit and Bray, 1989), while Cooper and Bauck (1989) recommend that artificial heads be equalized flat for free-field sounds at ±30°.

To provide a net flat response from microphone to listener, the artificial-head equalization should be the inverse of the headphone frequency response. If the head is equalized with a dip around 3 kHz to yield a net flat response, the headphones should have a mirror-image peak around 3 kHz (most do).

Artificial-Head Imaging with Loudspeakers

How does an artificial-head recording sound when reproduced over loudspeakers? According to Griesinger (1989), it can sound just as good as an ordinary stereo recording, with superior reproduction of location, height, depth, and hall ambience. But it sounds even better over headphones. Images in binaural recordings are mainly up front when you listen with speakers but are all around when you listen with headphones.

Genuit and Bray (1989) report that more reverberation is heard over speakers than over headphones, due to a phenomenon called *binaural reverberance suppression*. For this reason, it is important to monitor artificial-head recordings with headphones and speakers.

Griesinger notes that a dummy head must be placed relatively close to the musical ensemble to yield an adequate ratio of direct-to-reverberant sound over loudspeakers. This placement yields exaggerated stereo separation with a hole in the middle. However, the center image can be made more solid by boosting in the presence range (see Griesinger's, 1989, recommended equalization previously).

Although a dummy-head binaural recording can provide excellent imaging over headphones, it produces inadequate spaciousness at low frequencies over loudspeakers (Huggonet and Jouhaneau, 1987) unless spatial equalization is used (Griesinger, 1989). Spatial equalization was discussed in Appendix B under the heading "Coincident Systems with Spatial Equalization (Shuffler Circuit)." A low-frequency boost in the $L - R$ difference signal of about 15 dB at 40 Hz and ±1 dB at 400 Hz can improve the low-frequency separation over speakers.

References

Cooper, D. and Bauck, J. "Prospects for Transaural Recording." *Journal of the Audio Engineering Society*, Vol. 37, No. 1/2 (January–February 1989), pp. 3–19.

Geil, F. "Experiments with Binaural Recording." *db* (June 1979), pp. 30–35.

Genuit, K. and Bray, W. "The Aachen Head System: Binaural Recording for Headphones and Speakers." *Audio* (December 1989), pp. 58–66.

Griesinger, D. "Equalization and Spatial Equalization of Dummy Head Recordings for Loudspeaker Reproduction." *Journal of the Audio Engineering Society*, Vol. 37, No. 1/2 (January–February 1989), pp. 20–29.

Huggonet, C. and Jouhaneau, J. "Comparative Spatial Transfer Function of Six Different Stereophonic Systems." Preprint No. 2465 (H5), Paper Presented at the *Audio Engineering Society 82nd Convention*, March 10–13, 1987, London, p. 16, Fig. 13.

Moller, H. "Reproduction of Artificial-Head Recordings through Loudspeakers." *Journal of the Audio Engineering Society*, Vol. 37, No. 1/2 (January–February 1989), pp. 30–33.

Peus, S. "Development of a New Studio Artificial Head." *db Magazine* (June 1989), pp. 34–36.

Sunier, J. "A History of Binaural Sound." *Audio* (March 1989a), pp. 312–346.

Sunier, J. "Binaural Overview: Ears Where the Mics Are, Part 1." *Audio* (November 1989b), pp. 75–84.

Sunier, J. "Binaural Overview: Ears Where the Mics Are, Part 2." *Audio* (December 1989c), pp. 48–57.

Several papers on binaural sound were presented at the *89th Convention of the Audio Engineering Society*, September 21–25, 1990, Los Angeles. These papers are:

"Subjective Evaluation of Spatial Image Formation Processors," Elizabeth A. Cohen and Charles M. Salter Associates, Inc., San Francisco, CA.

"A New Method for Spatial Enhancement in Stereo and Surround Recording," Dr. Wieslaw R. Woszczyk, McGill University, Montreal, Canada.

"Multi-Channel Sound in the Home: Further Developments of Stereo-phony," Gunther Theile, Institut fur Rundfunktechnik, GmbH.

"Development and Use of Binaural Recording Technology," W. Bray, K. Genuit, and H. W. Gierlich, Jaffe Acoustics, Norwalk, CT.

"Spaciousness Enhancement of Stereo Reproduction Using Spectral Stereo Techniques," D. J. Furlong and A. G. Garvey, Preprint 3007.

"An Intuitive View of Coincident Stereo Microphones," S. Julstrom, Preprint 2984.

More-recent Audio Engineering Society preprints:

"Investigations on a New Reproduction Procedure for Binaural Recordings," Ning Xiang, Klaus Genuit, and Hans W. Gierlich, Head Acoustics, Herzogenrath, Germany, #3732, October 1993.

"Temporal Localization Cues and Their Role in Auditory Perception," Martin D. Wilde, Wilde Acoustics, Chicago, IL, #3708, October 1993.

"Early Reflections and Reverberant Field Distribution in Dual Microphone Stereophonic Sound Recording Systems," Michael Williams, Paris, France, #3155 (R4), October 1991.

"Binaural Record/Reproduction Systems and Their Use in Psychoacoustic Investigations," Floyd E. Toole, National Research Council Canada, Ottawa, Ontario, #3179 (L6), October 1991.

"Development and Use of Binaural Recording Techniques," K. Genuit, H. W. Gierlich, and Wade Bray, HEAD Acoustics, Aachen, Germany, Norwalk, CT, #2950, September 1990.

"Further Developments of Loudspeaker Stereophony," Gunther Theile, Institut fur Rundfunktechnik GmbH, Munich, Germany, #2947, September 1990.

"Microphone Arrays Optimized for Music Recording," W. Woszczyk, McGill University, #3255, March 1992.

"Frequency Dependent Hybrid Microphone Arrays for Stereophonic Sound Recording," Michael Williams, Paris France, #3252, March 1992.

"Standard Stereo Recording Techniques in Non-Standard Situations," Albert G. Swanson, Location Recording, Seattle, #3313, March 1992.

"Improved Externalization and Frontal Perception of Headphone Signals," Soren Gert Weinrich, Oticon A/S Research Unit, Snekkersten, Denmark, #3291, March 1992.

"BAP Binaural Audio Processor," F. Richter, AKG Acoustics, #3323, March 1992.

"Transfer Characteristics of Headphones," Henrik Moller et al., Institute for Electronic Systems, #3290, March 1992.

"Improved Possibilities of Binaural Recording and Playback Techniques," K. Genuit et al., HEAD Acoustics, Herzogenrath, Germany, #3332, March 1992.

"Applications of Blumlein Shuffling to Stereo Microphone Techniques," Michael Gerzon, Oxford, UK, #3448 (S-1), October 1992.

Preprints can be ordered from the Audio Engineering Society, www.aes.org

More articles in the *Journal of the Audio Engineering Society*:

"Measuring a Dummy Head in Search of Pinna Cues," H. L. Han, January–February 1994.

"Binaural Technique: Do We Need Individual Recordings?" Henrik Moller et al., Acoustics Laboratory, Aalborg University, Aalborg, Denmark, June 1996.

"Comments on 'Spaciousness and Localization in Listening Rooms and Their Effects on the Recording Technique'," Stanley Lipshitz, Audio Research Group, University of Waterloo, Waterloo, Ont., Canada, December 1987.

"The Effect of Head Shape on Spectral Stereo Theory," K. Rasmussen and P. Juhl, Acoustics Laboratory, Technical University of Denmark, Denmark, March 1993.

"On the Naturalness of Two-Channel Stereo Sound," Gunther Theile, Institut fur Rundfunktechnik GmbH, Munich, Germany, October 1991.

"A Computer Model of Binaural Localization for Stereo Imaging Measurement," E. Macpherson, Audio Research Group, University of Waterloo, Waterloo, Ont., Canada, September 1991.

"Room-Related Balancing Technique: A Method for Optimizing Recording Quality," M. Wohr, G. Theile, H. Goeres, and A. Persterer, September 1991.

ACKNOWLEDGMENTS

Thank you to all the manufacturers who sent photographs for this book. I greatly appreciate the contributions and advice of reviewers Jim Loomis, an instructor at Ithaca College; Bruce Outwin, an instructor at Emerson College; and Ron Estes of NBC in Burbank. A special thanks to Sam Kambol for his clever ideas. Thanks very much to Catharine Steers, Stephanie Barrett and Lyndsey Dixon at Focal Press for their skillful editoral assistance.

Thanks also to Paul Gottehrer at Focal Press for his fine coordination efforts.

Thank you to the publishers who allowed me to use some of my own material for this book. Some chapters are reprinted with permission from:

MR&M Publishing Corp. and Sagamore Publishing Co. Inc., "Recording Techniques" series by Bruce Bartlett.

Radio World, "Stereo Microphone Techniques Part 1," November 1989, and "Stereo Microphone Techniques Part 2," February 1990, by Bruce Bartlett.

Parts of Appendix C were based on the article by B. Bartlett, "An Improved Stereo Microphone Array Using Boundary Technology: Theoretical Aspects," *Journal of the Audio Engineering Society*, Vol. 38, No. 7/8, (July/August 1990), pp. 543–552.

GLOSSARY

A–B   *See* Spaced-pair method. Also, an A–B test is a listening comparison between two audio programs, or between two components playing the same program, performed by switching immediately from one to the other. The levels of the two signals are matched.

Accent microphone   *See* Spot microphone.

Ambience   Room acoustics, early reflections, and reverberation. Also, the audible sense of a room or environment surrounding a recorded instrument.

Ambience microphone   A microphone placed relatively far from its sound source to pick up ambience.

Amplitude   Level, intensity, or magnitude. For a sine wave, the rms (root mean square) amplitude of a sound wave or signal is almost the same as the average amplitude. The rms value is the effective or DC equivalent value. The peak amplitude is the voltage of the signal waveform peak. The rms amplitude is 0.707 times the peak amplitude.

Analog-to-digital (A/D) converter   A circuit that converts an analog signal to a digital signal.

Antiphase   Referring to two identical signals in opposite polarity. *See* Polarity.

Artificial head   *See* Dummy head.

Assign   To route or send an audio signal to one or more selected channels.

Attenuate   To reduce the level of a signal.

Audio interface (audio I/O box)   A device with mic and line input connectors, and a USB or FireWire output connector, that converts analog audio to digital and sends it to a computer USB or FireWire port for recording.

Aux bus   In a mixing console, the bus that feeds effects devices (signal processors) or a monitor power amplifier. A submixer in a mixing console that combines signals from aux sends, and then feeds the mixed signal to the input of an effects device.

Aux return   In the output section of a mixing console, a control that adjusts the amount of signal received from an effects unit. Also, the connectors in a mixer to which you connect the effects-unit output signal. They might be labeled "bus in" instead. The effects-return signal is mixed with the program bus signal.

Aux send   In each input module of a mixing console, a control that adjusts the amount of signal sent to an effects unit or monitor power amplifier. Also, the connectors in a mixer to which you connect the effects-unit input signal.

Baffled-omni   A stereo miking arrangement that uses two ear-spaced omnidirectional microphones separated by a hard padded baffle.

Balance   The relative volume levels of tracks in a mix or instruments in a musical ensemble.

Balanced line   A cable with two conductors surrounded by a shield, in which each conductor is at equal impedance to ground. With respect to ground, the conductors are at equal potential but opposite polarity; the signal flows through both conductors.

Binaural recording   A two-channel recording made with an omnidirectional microphone mounted in or near each ear of a human or a dummy head, for playback over headphones. The object is to duplicate the acoustic signal appearing at each ear.

Blumlein array   A stereo microphone technique in which two coincident bidirectional microphones are angled 90° apart (45° to the left and right of center).

Board   *See* Mixing console.

Boundary microphone   A microphone designed to be used on a boundary (a hard reflective surface). The microphone capsule is mounted very close to the boundary (or flush with it), so that direct and reflected sounds arrive at the microphone diaphragm in phase (or nearly so) for all frequencies in the audible band.

Bus   A common connection of many different signals. An output of a mixer or submixer. A channel that feeds a recorder track, signal processor, or power amplifier.

Bus in   An input to a program bus, usually used for effects returns.

Bus master   In the output section of a mixing console, a potentiometer (fader or volume control) that controls the output level of a bus.

Bus out   The output connector of a bus.

Buzz   An unwanted edgy tone that sometimes accompanies audio, containing high harmonics of 60 Hz (50 Hz in the UK and Europe).

CardBus   A faster version of a PC card, a CardBus card supports computer-bus mastering, speeds up to 33 MHz, and 32 bits instead of 16 bits. *See* PCMCIA.

Cardioid microphone   A unidirectional microphone with side attenuation of 6 dB and maximum rejection of sound at the rear of the microphone (180° off axis). A microphone with a heart-shaped directional pattern.

Channel   A single path of an audio signal. Usually, each channel contains a different signal.

Channel assign   *See* Assign.

Clean   Free of noise, distortion, overhang, and leakage. Not muddy.

Clear   Easy to hear, easy to differentiate. Reproduced with sufficient high frequencies and not too much reverberation.

Clip   (1) *See* Region. (2) To turn up an audio signal so high that the peaks of the audio waveform are clipped off, causing distortion. (3) A clip LED in a mixer flashes to indicate signal clipping.

Closely spaced method   *See* Near coincident.

Coincident-pair method   A stereo microphone, or two separate microphones, placed so that the microphone diaphragms occupy approximately the same point in space. They are angled apart and mounted one directly above the other.

Comb-filter effect   The frequency response caused by combining a sound with its delayed replica. The frequency response has a series of peaks and dips caused by phase interference. The peaks and dips resemble the teeth of a comb. This effect can occur with near-coincident and spaced-pair techniques when the left- and right-channel signals are combined to mono.

Compact flash card   A type of flash-memory card used to store data, such as digital audio recorded by a portable flash-memory recorder. *See also* Flash memory.

Compressor   A signal processor that reduces dynamic range by means of automatic volume control; an amplifier whose gain decreases as the input-signal level increases above a preset point.

Condenser microphone   A microphone that works on the principle of variable capacitance to generate an electrical signal. The microphone diaphragm and an adjacent metallic disk (called a backplate) are charged to form two plates of a capacitor. Incoming sound waves vibrate the diaphragm, varying its spacing to the backplate, which varies the capacitance, which in turn varies the voltage between the diaphragm and the backplate.

Connector   A device that makes electrical contact between a signal-carrying cable and an electronic device, or between two cables. A device used to connect or hold together a cable and an electronic component so that a signal can flow from one to the other.

Console   *See* Mixing console.

Contact pickup   *See* Pickup.

Convolution reverb (sampling reverb)   A reverberation device or plug-in that creates the reverb from impulse-response samples (wave files) of real acoustic spaces, rather than from algorithms. The resulting sound quality is very natural.

Crosstalk   The unwanted transfer of a signal from one channel to another. Head-related crosstalk is the right-speaker signal that reaches the left ear, and the left-speaker signal that reaches the right ear. In the transaural stereo system, this acoustic crosstalk is canceled by processing the stereo signal with electronic crosstalk that is the inverse of the acoustic crosstalk.

DAT (R-DAT)   A digital audio tape recorder that uses a rotating head to record digital audio on tape.

DAW   Abbreviation for digital audio workstation. *See* Digital audio workstation.

dB   Abbreviation for decibel. *See* Decibel.

dBA   Decibels, A weighted.

dBm   Decibels relative to 1 milliwatt.

dBu   Decibels relative to 0.775 volt.

dBV   Decibels relative to 1 volt.

Dead   Having very little or no reverberation.

Decay   The portion of the envelope of a note in which the envelope goes from maximum to some midrange level. Also, the decline in level of reverberation over time.

Decay time   *See* Reverberation time.

Decibel   The unit of measurement of audio level. Ten times the logarithm of the ratio of two power levels. Twenty times the logarithm of the ratio of two voltages. A decibel measurement is relative to some other level, not an absolute measurement. *See* dB.

Delay   The time interval between a signal and its repetition. A digital delay is a signal processor or plug-in that delays a signal for a short time.

Depth   The audible sense of nearness and farness of various instruments. Instruments recorded with a high ratio of direct-to-reverberant sound are perceived as being close; instruments recorded with a low ratio of direct-to-reverberant sound are perceived as being distant.

Design center   The portion of fader travel (usually shaded), about 10–15 dB from the top, in which console gain is distributed for optimum headroom and signal-to-noise ratio. During normal operation, master faders and group faders should be placed at or near design center.

Designation strip   A strip of tape attached near console faders to designate the instrument that each fader controls.

Desk   The British term for mixing console.

Destructive editing   In a digital audio workstation, editing that rewrites the data on disk. A destructive edit cannot be undone unless a backup copy was made of the data before it was written over.

DI   Acronym for direct injection, recording with a direct box.

Diffuse field   A sound field in which the sounds arrive randomly from all directions, such as the reverberant field in a concert hall. Diffuse-field equalization might be applied to a dummy head so that it has a net flat response in a diffuse sound field.

Digital audio   Encoding an analog audio signal in the form of binary digits (ones and zeros).

Digital audio workstation (DAW)   A system, device, or software that allows you to record, edit, and mix audio programs entirely in digital form.

A computer DAW includes a computer, recording software, and sound card or audio interface. It has virtual controls on-screen. A stand-alone DAW is a digital recorder-mixer with real mixer controls.

Digital recording   A recording system in which the audio signal is stored in the form of binary digits (ones and zeros).

Digital-to-analog converter   A circuit that converts a digital audio signal into an analog audio signal.

DIN   A German standard for a near-coincident stereo microphone technique in which two cardioid microphones are angled apart 90° and spaced 20 cm (7.9 in) horizontally.

Direct box   A device used for connecting an electric musical instrument directly to a mixer mic input. The direct box converts a high-impedance unbalanced audio signal into a low-impedance balanced audio signal.

Direct injection (DI)   Recording with a direct box.

Directional microphone   A microphone that has different sensitivity in different directions. A unidirectional or bidirectional microphone.

Direct out   In a mixing console, an output connector following a mic preamplifier, fader, and equalizer, used to feed the signal of one instrument to one track of a multitrack recorder. *See also* Insert jack.

Direct sound   Sound traveling directly from the sound source to the microphone (or to the listener) without reflections.

Distortion   An unwanted change in the audio waveform, causing a raspy or gritty sound quality. The appearance of frequencies in a device's output signal that were not in the input signal. Distortion is caused by recording at too high a level, improper mixer settings, components failing, or overdriving an amplifier. (Distortion can be desirable—for an electric guitar, for example.)

Dry   Having no echo or reverberation. Referring to a close-sounding signal that is not yet processed by an effects device or plug-in.

DSD   Abbreviation for direct stream digital, a Sony trademark for 1-bit encoding of digital signals used in their Super Audio CD format.

DSP   Abbreviation for digital signal processing, modifying a signal in digital form.

Dummy head   A modeled head with microphones in the ears, used for binaural recording; same as artificial head.

DVD   Digital versatile disc. A storage medium the size of a compact disc that holds much more data. The DVD stores video, audio, or computer data.

Dynamic microphone   A microphone that generates electricity when sound waves cause a conductor to vibrate in a stationary magnetic field. The two types of dynamic microphone are moving coil and ribbon. A moving-coil microphone is usually called a dynamic microphone. *See* Moving-coil microphone.

Dynamic range   The range of volume levels in a program from softest to loudest.

Earth ground   A connection to moist dirt (the ground we walk on). This connection is usually done via a long copper rod.

Echo   A delayed repetition of a signal or a sound. A sound delayed 50 milliseconds or more, combined with the original sound.

Editing   In a DAW, the cutting and rejoining of an audio waveform to delete unwanted material, to insert silence, or to rearrange recorded material into the desired sequence.

Effects   Interesting sound phenomena created by signal processors or plug-ins, such as reverberation, echo, flanging, doubling, compression, or chorus.

Effects loop   A set of connectors in a mixer for connecting an external effects unit, such as a reverb or delay device. The effects loop includes a send section and a return section. *See* Aux send, Aux return.

Electret-condenser microphone   A condenser microphone in which the electrostatic field of the capacitor is generated by an electret—a material that permanently stores an electrostatic charge.

Electrostatic field   The force field between two conductors charged with static electricity.

Electrostatic interference   The unwanted presence of an electrostatic hum field in signal conductors.

Elevation   An image displacement in height above the speaker plane.

End-addressed   Referring to a microphone whose main axis of pickup is perpendicular to the front of the microphone. You aim the front of the mic at the sound source. *See* Side-addressed.

Envelope   The rise and fall in volume of one note. The envelope connects successive peaks of the waves comprising a note. Each harmonic in the note might have a different envelope.

Equalization (EQ)   The adjustment of frequency response to alter the tonal balance or to attenuate unwanted frequencies.

Equalizer   A circuit and its controls (usually in each input module of a mixing console or in a separate unit) that alter the frequency spectrum of a signal passed through it.

Fade-out   To gradually reduce the volume of the last several seconds of a recorded song, from full level down to silence, by slowly pulling down the master fader.

Fader   A linear or sliding potentiometer (volume control), used to adjust signal level.

Faulkner method   Named after Tony Faulkner, a stereo microphone technique using two bidirectional microphones aiming at the sound source and spaced about 8 inches apart.

Feed   (1) To send an audio signal to some device or system. (2) An output signal sent to some device or system.

Feedback   (1) The return of some portion of an output signal to the system's input. (2) The squealing sound you hear when a PA system microphone picks up its own amplified signal through a loudspeaker.

Filter   A circuit that sharply attenuates frequencies above or below a certain frequency. Used to reduce noise and leakage above or below the frequency range of an instrument or voice.

FireWire   A standard protocol and port for high-speed transfer of data between digital devices. Also called IEEE 1394. Connects a computer to external devices such as MIDI interfaces, memory sticks, memory recorders, and audio interfaces. Faster than a standard serial port.

Flash memory   An erasable, removable computer memory card that can store data permanently.

Float   To disconnect from ground.

Focus   The degree of fusion, compactness, or positional definition of a sonic image. A sharp image is highly focused and easy to locate.

FOH Abbreviation for front-of-house, the location of the sound-reinforcement mixer in a venue.

Free field The sound field coming directly from the sound source without reflections; the sound field in an anechoic chamber. Free-field equalization can be applied to a dummy head to make it have a net flat response in a free field.

Frequency The number of cycles per second of a sound wave or an audio signal, measured in hertz (Hz). A low frequency (for example, 100 Hz) has a low pitch; and a high frequency (for example, 10,000 Hz) has a high pitch.

Frequency response (1) The range of frequencies that an audio device will reproduce at an equal level (within a tolerance, such as ±3 dB). (2) The range of frequencies that a device (mic, human ear, etc.) can detect.

Fundamental The lowest frequency in a complex wave.

Fusion The formation of a single image by two or more sound sources, such as loudspeakers.

Gain Amplification. The ratio, expressed in decibels, between the output voltage and the input voltage, or between the output power and the input power.

Graphic equalizer An equalizer with a horizontal row of faders; the fader-knob positions indicate graphically the frequency response of the equalizer. Usually used to equalize monitor speakers for the room they are in. Sometimes used for complex EQ of a track.

Ground The zero-signal reference point for a system of audio components.

Ground bus A common connection to which equipment is grounded, usually a heavy copper plate.

Grounding Connecting pieces of electronic equipment to ground. Proper grounding ensures that there is no voltage difference between equipment chassis. An electrostatic shield needs to be grounded to be effective.

Ground loop (1) A loop or circuit formed of ground leads. (2) The loop formed when two or more audio components are connected together via two ground paths: the connecting-cable shield and the safety ground pins in the components' AC power cords. If there is a voltage difference between the two equipment grounds, the ground loop causes hum.

Group *See* Submix.

Hard disk   A random-access storage medium for computer data. A hard disk drive contains a stack of magnetically coated hard disks that are read by, and written to by, an electromagnetic head.

Hard disk recorder (HD recorder)   A device dedicated to recording digital audio on a hard disk drive. A hard disk recorder-mixer includes a built-in mixer.

Harmonic   In a complex wave, an overtone whose frequency is a whole-number multiple of the fundamental frequency.

Headroom   The safety margin, measured in decibels, between the signal level and the maximum undistorted signal level.

Hertz (Hz)   Cycles per second, the unit of measurement of frequency.

High-pass filter   A filter that passes frequencies above a certain frequency and attenuates frequencies below that same frequency. A low-cut filter.

Hiss   A noise signal containing all frequencies, but with greater energy at higher octaves. Hiss sounds like wind blowing through trees. It is usually caused by random signals generated by microphones and electronics.

Hot   (1) A high recording level causing slight distortion, may be used for special effect. (2) A condition in which a chassis or circuit has a potentially dangerous voltage on it. (3) Referring to the conductor in a microphone cable which has a positive voltage on it at the instant that sound pressure moves the diaphragm inward.

Hum   An unwanted low-pitched tone (60 Hz and its harmonics, 50 Hz in Europe) heard in the monitors. The sound of interference generated in audio circuits and cables by AC power wiring. Hum pickup is caused by such things as faulty grounding, poor shielding, and ground loops.

Hypercardioid microphone   A directional microphone with a polar pattern that has 12 dB attenuation at the sides, 6 dB attenuation at the rear, and two nulls of maximum rejection at 110° off axis.

Image   An illusory sound source located somewhere around the listener. An image is generated by two or more loudspeakers. In a typical stereo system, images are located between the two stereo speakers.

Imaging   The ability of a microphone array or a speaker pair to form easily localizable images.

Impedance   The opposition of a circuit to the flow of alternating current, measured in ohms. Impedance is the complex sum of resistance and reactance. Abbreviated as Z. In a microphone, low impedance is under 600 ohms, medium impedance is about 1500 ohms, and high impedance is over 10 kilohms.

Input   The connection going into an audio device. In a mixer or mixing console, a connector for a microphone, line-level device, or other signal source.

Input attenuator   *See* Attenuator.

Input module   In a mixing console, the set of controls affecting a single input signal. An input module usually include an attenuator (trim), fader, equalizer, aux sends, and channel-assign controls.

Input section   The row of input modules in a mixing console.

Insert jacks (insert sockets outside the US)   One or two jacks (sockets outside the US) in a console input module or output module that allow access to points in the signal path, usually for connecting a compressor. Plugging into the insert connectors breaks the signal flow and allows you to insert a signal processor or multitrack recorder in series with the signal. The insert-send connector can feed a signal to a recorder track input, while the insert-return connector can accept a signal from a recorder's track output.

Intensity stereo (XY stereo)   A method of forming stereo images by intensity or amplitude differences between channels. *See* Coincident-pair method.

ITE/PAR   Acronym for In the Ear/Pinna Acoustic Response, a stereo recording system developed by Don and Carolyn Davis of Synergetic Audio Concepts. It uses two probe microphones in the ear canals, near the ear drum of a human listener. Playback is over two speakers up front and two to the sides of the listener.

Jack (US definition)   A female or receptacle-type connector for audio signals into which a plug is inserted. Outside the US, a jack is called a socket, and a jack plug inserts into a socket.

Jecklin disk   Named after its inventor, a stereo microphone array using two omnidirectional microphones spaced 6½ inches apart and separated by a disk or baffle 11⅞ inches in diameter, covered with flat sound absorbent material; also known as the OSS system. *See also* Schneider disk.

Kilo   A prefix meaning one thousand, abbreviated k.

Leakage   The overlap of an instrument's sound into another instrument's microphone. Also called bleed or spill.

Level   The degree of intensity of an audio signal; the voltage, power, or sound pressure level. The original definition of level is the power in watts.

Level setting   In a recording system, the process of adjusting the input-signal level to obtain maximum level on the recording media without distortion. A meter shows recording level.

Limiter   An amplifier whose output is constant above a preset input level. A compressor with a compression ratio of 10:1 or greater, with the threshold set just below the point of distortion of the following device. Used to prevent distortion of attack transients or peaks.

Line level   In balanced professional recording equipment, a signal whose nominal level is approximately 1.23 volts (+4 dBu). In unbalanced equipment (most home hi-fi or semipro recording equipment), a signal whose level is approximately 0.316 volt (−10 dBV).

Live   (1) Having audible reverberation. (2) Occurring in real time, in person.

Live recording   A recording made at a gig or concert. Also, a recording made of a musical ensemble playing all at once, rather than overdubbing.

Localization   Our ability to tell the direction of a real sound source or an image (illusory sound source).

Localization accuracy   The accuracy with which a stereo microphone array translates the location of real sound sources into image locations. If localization is accurate, instruments at the side of the musical ensemble are reproduced from the left or right speaker; instruments halfway off-center are reproduced halfway between the center of the speaker pair and one speaker, and so on.

Location   The angular position of an image relative to a point straight ahead of a listener, or its position relative to the loudspeakers.

M   Abbreviation for mega, or one million (as in megabytes).

Master   (1) A completed CD used to generate compact discs. (2) To master an audio program is to put the song mixes in order, insert silent spaces between them, and match their volumes and tonal balances.

Master fader A volume control that affects the level of all program buses simultaneously. It is the last stage of gain adjustment before the two-track recorder.

Memory recorder A device that records two tracks of audio to a flash-memory card in MP3 or wave format.

Meter A device that indicates voltage, resistance, current, or signal level.

Mic An abbreviation for microphone.

Mic level The level or voltage of a signal produced by a microphone, typically 2 millivolts.

Mic preamp *See* Preamplifier.

Microphone A transducer or device that converts an acoustical signal (sound) into a corresponding electrical signal.

Microphone techniques The selection and placement of microphones to pick up sound sources.

Mid–side A coincident-pair stereo microphone technique using a forward-facing unidirectional, omnidirectional, or bidirectional mic and a side-facing bidirectional mic. The microphone signals are summed and differenced to produce right- and left-channel signals.

Mike To pick up with a microphone.

Milli A prefix meaning one-thousandth, abbreviated m.

Mix (1) To combine two or more different signals into a common signal. (2) A control on an effect unit that varies the ratio between the dry and the processed signals.

Mixdown The process of playing recorded tracks through a mixing console and mixing them to two stereo channels or six surround channels.

Mixer A device that mixes or combines audio signals and controls the relative levels of the signals.

Mixing console A large mixer with additional functions such as equalization or tone control, pan pots, monitoring controls, solo functions, channel assigns, and aux sends.

Monaural Referring to listening with one ear; often incorrectly used to mean monophonic.

Monitor   To listen to an audio signal with headphones or loudspeakers. Also, a loudspeaker in a control room, or headphones, used for judging sound quality. Also, a video display screen used with a computer.

Monitoring   Listening to an audio signal with a monitor.

Mono, monophonic   Referring to a single channel of audio. A monophonic program can be played over one or more loudspeakers, or one or more headphones.

Mono-compatible   A characteristic of a stereo program, in which the program channels can be combined to a mono program without altering the frequency response or balance. A mono-compatible stereo program has the same frequency response in stereo or mono because there is no delay or phase shift between channels to cause phase interference.

Moving-coil microphone   A microphone with a diaphragm attached to a coil of wire moving in a magnetic field. The diaphragm vibrates when struck with sound waves, which vibrates the coil and generates an electrical signal similar to the incoming sound wave. Usually called a dynamic microphone.

MP3 (MPEG Level-1 Layer-3)   A data compression format for audio. In an MP3 file (.mp3), the data has been compressed or reduced typically to 1/10 its original size or less. Compressed files take up less memory, so they download faster. You download MP3 files to your hard drive, then listen to them. MP3 audio quality at a 128 kbps rate is nearly the same as that of CDs (depending on source material). *See also* WMA.

MS recording   *See* Mid–side.

Muddy   Unclear sounding; having excessive leakage, reverberation, or an undamped envelope.

Multitrack   Referring to a recorder that has more than two tracks.

Mute   To turn off an input signal on a mixing console by disconnecting the input-module output from channel assign and direct out. During mixdown, the mute function in a track is used to reduce noises and leakage during silent portions of the audio, or to turn off unused performances. During recording, mute is used to turn off mic signals.

Near coincident   A stereo microphone technique in which two directional microphones are angled apart symmetrically on either side of center and spaced a few inches apart horizontally.

Nearfield™ monitoring   A monitor-speaker arrangement in which the speakers are placed about 3 feet apart and 3 feet from the listener to reduce the audibility of control-room acoustics.

Noise   Unwanted sound, such as hiss from electronics or tape. An audio signal with an irregular, non-periodic waveform.

Non-destructive editing   In a digital audio workstation, editing done by changing pointers (location markers) to information on the hard disk. A non-destructive edit can be undone.

Nonlinear   (1) Referring to a storage medium in which any data can be accessed or read almost instantly. Examples are a hard disk, compact disc, and MiniDisc. *See* Random access. (2) Referring to an audio device that is distorting the signal.

NOS   A Dutch Broadcasting System standard for a near-coincident stereo microphone technique in which two cardioid microphones are angled apart 90° and spaced 30 cm (11.8 in) horizontally.

Off axis   Not directly in front of a microphone or a loudspeaker.

Off-axis coloration   In most microphones, the deviation from the on-axis frequency response that occurs at angles off the axis of the microphone. The coloration of sound (alteration of tone quality) for sounds arriving off axis to the microphone.

Omnidirectional microphone   A microphone that is equally sensitive to sounds arriving from all directions.

On-location recording   A recording made outside the studio, in a room or hall where the music usually is performed or practiced.

ORTF   Named after the French broadcasting network (Office de Radio-diffusion Television Française), a near-coincident stereo mic technique that uses two cardioid mics angled 110° apart and spaced 17 cm horizontally.

OSS system   Abbreviation for optimal stereo signal system. *See* Jecklin disk.

Output   A connector in an audio device from which the signal comes and feeds successive devices.

Overdub   To record a new musical part on an unused track in synchronization with previously recorded tracks.

275

Overload   The distortion that occurs when an applied signal exceeds a system's maximum input level.

Pad   A resistive network that reduces the microphone signal level to prevent overloading of the input transformer and mic preamplifier.

Pan pot   Abbreviation for panoramic potentiometer. In each input module in a mixing console, a control that divides a signal between two channels in an adjustable ratio. By doing so, a pan pot controls the location of a sonic image between a stereo pair of loudspeakers.

PC card   *See* PCMCIA.

PCMCIA   An acronym for Personal Computer Memory Card International Association, a standard for credit-card-size PC Cards that plug into portable computers to add extra functions such as memory, modems, portable disk drives, or USB/FireWire ports. Comes in three types that have different thicknesses.

Perspective   In the reproduction of a recording, the audible sense of distance to the musical ensemble, the point of view. A close perspective has a high ratio of direct sound to reverberant sound; a distant perspective has a low ratio of direct sound to reverberant sound.

PFL   Abbreviation for prefader listen. *See also* Solo.

Phantom image   *See* Image.

Phantom power   A DC voltage (usually 12–48 volts) applied to microphone signal conductors to power-condenser microphones.

Phantom power supply   A stand-alone device, or a circuit built into a mixer or mic preamp, that provides phantom power.

Phase   The degree of progression in the cycle of a wave, where one complete cycle is 360°.

Phase cancellation, phase interference   The cancellation of certain frequency components of a signal that occurs when the signal is combined with its delayed replica. At certain frequencies, the direct and delayed signals are of equal level and opposite polarity (180° out of phase), and when combined, they cancel out. The result is a comb-filter frequency response having a periodic series of peaks and dips. Phase interference can occur between the signals of two microphones picking up the same source at different distances, or can occur at a microphone picking up both a direct sound and its reflection from a nearby surface.

276

Phase shift   The difference in degrees of phase angle between corresponding points on two waves. If one wave is delayed with respect to another, there is a phase shift between them of $2\pi FT$, where $\pi = 3.14$, F = frequency in Hz, and T = delay in seconds.

Phone plug   (US definition) A cylindrical, coaxial connector of 1/4- or 1/8-inch diameter. An unbalanced phone plug has a tip for the hot signal and a sleeve for the shield or ground. A balanced phone plug has a tip for the signal hot signal, a ring for the return signal, and a sleeve for the shield or ground. A phone plug used with a TRS (tip–ring–sleeve) insert jack has a tip for the insert-send signal, a ring for the insert-return signal, and a sleeve for the common shield or ground.

Outside the US, a phone plug is called a jack plug. It is a cylindrical, coaxial connector of 6.35 or 3.5 mm diameter. An unbalanced jack plug has a tip for the hot signal and a sleeve for the shield or ground. A balanced jack plug has a tip for the signal hot signal, a ring for the return signal, and a sleeve for the shield or ground. A jack plug used with a TRS (tip–ring–sleeve) insert socket has a tip for the insert-send signal, a ring for the insert-return signal, and a sleeve for the common shield or ground.

Phono plug   A coaxial plug with a central pin for the hot signal and a ring of pressure-fit tabs for the shield or ground. Also called RCA plug.

Pickup   A piezoelectric transducer that converts mechanical vibrations to an electrical signal. Used in acoustic guitars, acoustic basses, and fiddles. Also, a magnetic transducer in an electric guitar that converts string vibration to a corresponding electrical signal. Same as contact pickup.

Pinnae   The outer ears. Reflections from folds of skin in the pinnae aid in localizing sounds.

Plug   A male connector that inserts into a jack outside the US, a jack plug is a male connector that inserts into a socket.

Plug-in power   DC voltage for a condenser microphone supplied from the connected recording equipment.

Plug-ins   Software effects that you install in your computer. The plug-in software becomes part of another program you are using (the host), such as a digital editing program. You can access the plug-in from the host software.

Polar pattern   The directional pickup pattern of a microphone. A plot of microphone sensitivity plotted versus angle of sound incidence. Examples of polar patterns are omnidirectional, bidirectional, and unidirectional. Subsets of unidirectional are cardioid, supercardioid, and hypercardioid.

Polarity   Referring to the positive or negative direction of an electrical, acoustical, or magnetic force. Two identical signals in opposite polarity are 180° out of phase with each other at all frequencies.

Pop   (1) A thump or little explosion sound heard in a vocalist's microphone signal. Pop occurs when the user says words with *p*, *t*, or *b*, so that a turbulent puff of air is forced from the mouth and strikes the microphone diaphragm. (2) A noise heard when a mic is plugged into a monitored channel or when a switch is flipped.

Pop filter   A screen placed on or near a microphone grille that attenuates or filters out pop disturbances before they strike the microphone diaphragm. Usually made of open-cell plastic foam or nylon fabric, a pop filter reduces pop and wind noise.

Power amplifier   An electronic device that amplifies or increases the power level fed into it to a level sufficient to drive a loudspeaker.

Power ground (safety ground)   A connection to the power company's earth ground through the U-shaped hole in a power outlet. In the power cable of an electronic component with a three-prong plug, the U-shaped prong is wired to the component's chassis. This wire conducts electricity to power ground if the chassis becomes electrically hot, preventing shocks.

Preamplifier (preamp)   In an audio system, the first stage of amplification that boosts a mic-level signal to line-level. A preamp is a stand-alone device or a circuit in a mixer.

Prefader–postfader switch   A switch that selects a signal either ahead of the fader (prefader) or following (postfader) the fader. The level of a prefader signal is independent of the fader position; the level of a postfader signal follows the fader position. Signals sent from mixer input modules to a multitrack recorder should be prefader so that fader adjustments don't change the recording level.

Preproduction   Planning in advance what will be done at a recording session, in terms of track assignments, overdubbing, instrument layout, and microphone selection.

Presence peak   A rise in the frequency response of a microphone around 5 kHz to add clarity or definition.

Pressure Zone Microphone   A boundary microphone constructed with the microphone diaphragm parallel to and facing a reflective surface.

Pressure-response microphone   An omnidirectional microphone whose frequency response is flat at high frequencies when the mic is used as a boundary microphone. *See* Boundary microphone.

Production   (1) A recording that is enhanced by effects. (2) The supervision of a recording session to create a satisfactory recording. This involves getting musicians together for the session, making musical suggestions to the musicians to enhance their performance, and making suggestions to the engineer for sound balance and effects.

Program bus   A bus or output that feeds an audio program to a recorder track.

Program mixer   In a mixing console, a mixer formed of input-module outputs, combining amplifiers, and program busses.

Proximity effect   The bass boost that occurs with a single-D directional microphone when it is placed a few inches from a sound source. The closer the microphone, the greater the low-frequency boost due to proximity effect.

Rack   A 19-inch-wide wooden or metal cabinet used to hold audio equipment.

Radio-frequency interference (RFI)   Radio-frequency electromagnetic waves induced in audio cables or equipment, causing various noises in the audio signal.

Random access   Referring to a storage medium in which any data point can be accessed or read almost instantly. Examples are a hard disk, compact disc, and MiniDisc.

R-DAT   *See* DAT.

Recorder-mixer   A combination of multitrack recorder and mixer in one chassis.

Reflected sound   Sound waves that reach the listener after being reflected from one or more surfaces.

Region   In a digital audio editing program, a defined segment of the audio program, such as a song, song section, musical phrase, or a note. Also called a clip.

Remote recording   *See* On-location recording.

Removable hard drive A hard disk drive that can be removed and replaced with another, used in a digital audio workstation to store audio and editing files, and used in some multitrack hard disk recorders to store digital audio.

Reverberation (reverb) Natural reverberation in a room is a series of multiple sound reflections that makes the original sound persist and gradually die away or decay. These reflections tell the ear that you're listening in a large or hard-surfaced room. For example, reverberation is the sound you hear just after you shout in an empty gymnasium. A reverb effect simulates the sound of a room—a club, auditorium, or concert hall—by generating random multiple echoes that are too numerous and rapid for the ear to resolve. The timing of the echoes is random, and the echoes increase in number with time as they decay. An echo is a discrete repetition of a sound; reverberation is a continuous fade-out of sound.

Reverberation time (RT60) The time it takes for reverberation to decay to 60 dB below the original steady-state level and has become inaudible.

RFI *See* Radio-frequency interference.

Ribbon microphone A dynamic microphone in which the conductor is a long metallic diaphragm (ribbon) suspended in a magnetic field. Usually a ribbon microphone has a bidirectional (figure-eight) polar pattern and can be used for the Blumlein method of stereo recording.

Sampling Recording a short sound event into computer memory. The audio signal is converted into digital data representing the signal waveform, and the data is stored in memory chips or on disk for later playback.

SASS The Stereo Ambient Sampling System™, a stereo microphone using two boundary microphones, each on a 5-inch square panel, angled apart and ear-spaced, with a baffle between the microphones.

Schneider disk Named after its inventor, a stereo microphone array using two omnidirectional microphones spaced 6½ inches apart and separated by a disk or baffle 11⅞ inches in diameter, covered with hemispheres of sound absorbent material. *See also* Jecklin disk.

SD (secure digital) card (compact flash card) A type of flash-memory card used to store data, such as digital audio recorded by a portable flash-memory recorder. Generally smaller physical size than a compact flash card. *See also* Flash memory.

Semi-coincident method *See* Near coincident.

Sensitivity   (1) The output of a microphone in volts for a given input in sound pressure level. (2) The sound pressure level a loudspeaker produces at 1 meter when driven with 1 watt of pink noise. *See also* Sound pressure level.

Shield   A conductive enclosure (usually a chassis, metal braid, or foil) around one or more signal conductors, used to keep out electrostatic fields that cause hum or buzz.

Shock mount   A suspension system that mechanically isolates a microphone from its stand or boom, preventing the transfer of mechanical vibrations.

Shotgun microphone (line microphone)   A highly directional microphone made of a slotted "line interference" tube mounted in front of a hypercardioid microphone capsule.

Shuffling   *See* Spatial equalization.

Sibilance   In a speech recording, excessive frequency components in the 5–10 kHz range, which are heard as an overemphasis of *s* and *sh* sounds.

Side-addressed   Refers to a microphone whose main axis of pickup is perpendicular to the side of the microphone. You aim the side of the mic at the sound source. *See also* End-addressed.

Signal   A varying electrical voltage that represents information, such as a sound.

Signal path   The path a signal travels from input to output in a piece of audio equipment.

Signal processor   A device that is used to alter a signal in a controlled way. Signal processors provide compression, gating, or equalization; or provide such effects as chorus, reverberation, pitch shift, and echo.

Signal-to-noise (S/N) ratio   The ratio in decibels between signal and noise voltage. An audio component with a high S/N has little background noise accompanying the signal; a component with a low S/N is noisy.

Single-D microphone   A directional microphone having a single distance between its front and rear sound entries. Such a microphone has proximity effect.

Size   *See* Focus.

SMPTE time code A modulated 1200 Hz square-wave signal used to synchronize two or more video transports with audio recorders. SMPTE is an abbreviation for the Society of Motion Picture and Television Engineers, which developed the time code. In Europe, engineers use the EBU time code (a 1000 Hz modulated square wave).

Snake A multipair or multichannel mic cable. Also, a multipair mic cable attached to a stage box containing mic connectors.

Solo On an input module in a mixing console, a switch that lets you monitor that particular input signal by itself. The switch routes only that input signal to the monitor system.

Sound card A circuit card that plugs into a PCI slot in a computer and converts an audio signal into computer data for storage in memory or on hard disk. The sound card also converts computer data into an audio signal.

Sound pressure level (SPL) The acoustic pressure of a sound wave, measured in decibels above the threshold of hearing. dB SPL $= 20 \log (P/P_{ref})$, where $P_{ref} = 0.0002$ dyne/cm^2.

Spaced-pair method A stereo microphone technique using two identical microphones spaced several feet apart horizontally, usually aiming straight ahead toward the sound source. Also called A–B.

Spatial equalization A low-frequency shelving boost in the $L - R$ (difference) signal of a stereo program, and a complementary shelving cut in the $L + R$ (sum) signal, in order to align the locations of the low- and high-frequency components of images, and to increase spaciousness or stereo separation.

Spatial processor A signal processor that allows images to be placed beyond the limits of a stereo pair of speakers, even behind the listener or toward the sides.

Spectrum The output versus frequency of a sound source, including the fundamental frequencies and harmonics.

SPL *See* Sound pressure level.

Splitter A transformer or circuit used to divide a microphone signal into two or more identical signals to feed different sound systems.

Spot microphone In classical music recording, a close-placed microphone that is mixed with more-distant microphones to add presence or to improve the balance.

Stage box   A chassis with several mic connectors, wired to a multiconductor cable called a snake. The stage box and snake carry the signals of several microphones to a mixer.

Stage width   *See* Stereo spread.

Stereo, stereophonic   An audio recording and reproduction system with correlated information between two channels (usually discrete channels), and meant to be heard over two or more loudspeakers to give the illusion of sound-source localization and depth. *Stereo* means "solid" or three-dimensional.

Stereo bar, stereo microphone adapter   A microphone stand adapter that mounts two microphones on a single stand for convenient stereo miking.

Stereo imaging   The ability of a stereo recording or reproduction system to form clearly defined audio images at various locations between a stereo pair of loudspeakers.

Stereo microphone   A microphone containing two mic capsules in a single housing for convenient stereo recording. The capsules usually are coincident.

Stereo spread   The reproduced stage width. The distance between the reproduced images of the left and right side of a musical ensemble.

Submaster (group master, bus master)   A master volume control for an output bus.

Submix   A small preset mix within a larger mix, such as a drum mix, keyboard mix, vocal mix, etc. Also a cue mix, monitor mix, or effects mix.

Submixer   A smaller mixer within a mixing console (or standing alone) that is used to set up a submix, a cue mix, an effects mix, or a monitor mix.

Super audio CD   Invented by Sony, a compact-disc format with two layers. One layer contains a two-channel DSD program followed by the same program in six channels for surround. The other layer contains a two-channel 16-bit/44.1 K linear PCM program for compatibility with existing compact-disc players.

Supercardioid microphone   A unidirectional microphone that attenuates side-arriving sounds by 8.7 dB, attenuates rear-arriving sounds by 11.4 dB, and has two nulls of maximum sound rejection at 125° off axis.

Surround microphone   A microphone with four or five microphones mounted on a common holder for convenient surround recording.

Surround sound A multichannel recording and reproduction system that plays sound all around the listener. The 5.1 surround system uses the following speakers: front-left, center, front-right, left-surround, right-surround, and subwoofer.

Sync, synchronization Aligning two separate audio programs in time, and maintenance of that alignment as the programs play.

Three-pin connector A 3-pin professional audio connector used for balanced signals. Pin 1 is soldered to the cable shield, pin 2 is soldered to the signal hot (in-polarity) lead, and pin 3 is soldered to the signal cold lead. *See also* XLR-type connector.

Three-to-one rule A rule in microphone applications. When multiple mics are mixed to the same channel, the distance between mics should be at least 3 times the distance from each mic to its sound source. This prevents audible phase interference.

Tie Connect electrically; for example, by soldering a wire between two points in a circuit.

Tight (1) Having very little leakage or room reflections in the sound pickup. (2) Refers to well-synchronized playing of musical instruments. (3) Having a well-damped, rapid decay.

Timbre The subjective impression of spectrum and envelope. The quality of a sound that allows us to differentiate it from other sounds. For example, if you listen to a trumpet, a piano, and a drum, you will hear that each has a different timbre or tone quality which identifies it as a particular instrument.

Time code *See* SMPTE time code.

Tonal balance The balance or volume relationships among different regions of the frequency spectrum, such as bass, midbass, midrange, upper midrange, and highs.

Track A group of bytes in a digital signal (on tape, on hard disk, on compact disc, or in a data stream) that represents a single channel of audio or MIDI. Usually one track contains a performance of one musical instrument.

Transaural stereo A method of recording surround sound heard over two loudspeakers. During recording, the signals from a dummy head are processed for playback over loudspeakers, so that acoustic crosstalk around the head is canceled. This crosstalk is the signal from the right speaker that reaches the left ear, and the signal from the left speaker that reaches the right ear. The net effect is to enable the listener to hear, over loudspeakers, what the dummy head heard in the original environment.

Transducer A device that converts energy from one form to another, such as a microphone or a loudspeaker.

Transformer An electronic component made of two magnetically coupled coils of wire. The input signal is transferred magnetically to the output, without a direct connection between input and output.

Transient A rapidly changing signal with a fast attack and a short decay, such as a drum beat.

Trim (1) In a mixing console, a control for fine adjustment of level, as in a bus trim control. (2) In a mixing console, a control that adjusts the gain of a mic preamp to accommodate various signal levels.

TRS Tip–ring–sleeve designation for a balanced (or stereo) phone plug or phone jack. Outside the US, TRS is the tip–ring–sleeve designation for a balanced (or stereo) jack plug or socket.

TS Tip–sleeve designation for an unbalanced (or mono) phone plug or phone jack. Outside the US, TS is the tip–sleeve designation for an unbalanced (or mono) jack plug or socket.

Tube A vacuum tube, an amplifying component made of electrodes in an evacuated glass tube. Tube sound is characterized as being "warmer" than solid state or transistor sound.

Unbalanced line An audio cable having one conductor surrounded by a shield that carries the return signal. The shield is at ground potential.

Unidirectional microphone A microphone that is most sensitive to sounds arriving from one direction, in front of the microphone. Examples are cardioid, supercardioid, and hypercardioid.

USB (Universal Serial Bus) A Mac/PC computer serial port and protocol for high-speed transfer of data between digital devices. Connects a computer to external devices such as MIDI interfaces, memory sticks, memory recorders, and audio interfaces. Faster than a standard serial port.

Valve The British term for vacuum tube.

Virtual controls Audio equipment controls that are simulated on a computer monitor screen. You adjust them with a mouse or a control surface.

Virtual loudspeaker   A transaural image synthesized to simulate a loud-speaker placed at a desired location.

Virtual surround system   An audio reproduction system using two speakers to create the illusion that the listener is surrounded by virtual loud-speakers in a 5.1 surround array.

Virtual track   A recording in a hard disk recorder-mixer of a single take of a performance. You choose which virtual track(s) that you want to feed to a real track during mixdown.

VU meter   A voltmeter with a specified transient response, calibrated in VU or volume units, used to show the relative volume of various audio signals, and to set recording level. In digital equipment, the VU meter is often replaced by the peak-reading LED or LCD meter.

Waveform   A graph of a signal's sound pressure or voltage versus time. The waveform of a pure tone is a sine wave.

Windscreen   A foam or fur cover for a microphone that rejects wind noise.

WMA (Windows Media Audio)   A popular compressed audio file format for streaming audio and for downloads. Windows Media 9.1 promises performance similar to that of MP3Pro: near-CD quality at 48 kbps and CD quality at 64 kbps.

XLR-type connector   An ITT Cannon part number that has become the popular definition for a 3-pin professional audio connector. *See also* Three-pin connector.

XY   *See* Coincident-pair method.

Y-adapter   A cable that divides into two cables in parallel to feed one signal to two destinations.

Z   Mathematical variable for impedance.

INDEX

CD LINER NOTES

Recording Music On Location
Recording Live Gigs and Concerts
Bruce and Jenny Bartlett

Welcome to "Recording Music On Location." This CD demonstrates various topics in the book.

All tracks were engineered by Bruce Bartlett. All music on this CD is copyrighted by the original artists and may not be reproduced without permission. Thanks to the Elwood Splinters Blues Band and Paul Thode for the use of their composition, "Water's Edge", copyright 1997.

This CD was produced with Cakewalk® SONAR Producer digital recording software. To provide the best sound quality, this CD was not processed to maximize its level.

Stereo Microphone Techniques

Speaker Setup
1. Introduction
2. Channel identification
3. Channel balance
4. Speaker polarity

Imaging Demo
5. Image location versus level differences between channels
6. Image location versus time differences between channels
7. Image location versus level and time differences between channels
8. **Coincident stereo mic techniques (intro)**
9. Coincident cardioids angled 90°
10. Coincident cardioids angled 120°
11. Coincident cardioids angled 180°
12. **Near-coincident mic techniques (intro)**
13. ORTF technique
14. NOS technique
15. **Baffled-omni techniques (intro)**
16. Jecklin disk
17. Sphere microphone
18. **Spaced-pair techniques (intro)**
19. Mics 2 feet apart
20. Mics 6 feet apart
21. **Boundary mic techniques (intro)**
22. Two PZMs spaced 3 feet apart
23. Two cardioid boundary mics with NOS technique

Miscellaneous
24. Drum set: spaced pair versus coincident pair
25. Miking distance and its perceived effect on depth

Popular Music Recording
26. Mixing a live recording of a blues band
27. Outro

Total running time 28:01.